Advances in
MARINE BIOLOGY
VOLUME 25

Advances in

MARINE
BIOLOGY

VOLUME 25

Edited by

J. H. S. BLAXTER
*Dunstaffnage Marine Research
Laboratory, Oban, Scotland*

and

A. J. SOUTHWARD
The Laboratory, Citadel Hill, Plymouth, England

Academic Press
Harcourt Brace Jovanovich, Publishers
London   San Diego   New York
Boston   Sydney   Tokyo   Toronto

ACADEMIC PRESS LIMITED
24/28 Oval Road
London NW1 7DX

United States Edition published by
ACADEMIC PRESS INC.
San Diego, CA 92101

British Library Cataloguing in Publication Data
Advances in marine biology.—Vol. 25
 1. Marine biology—Periodicals
 574.92'05   QH91.A1

ISBN 0-12-026125-1
ISSN 0065-2881

Typeset by Latimer Trend & Company Ltd, Plymouth, England
and printed in Great Britain
at T.J. Press (Padstow) Ltd, Cornwall

CONTRIBUTORS TO VOLUME 25

K. M. BAILEY, *Northwest and Alaska Fisheries Center, National Marine Fisheries Service, 7600 Sand Point Way NE, Seattle, Washington, 98115, USA.*

S. V. BOLETSKY, *Laboratoire Arago, C.N.R.S. and University of Paris 6 (U.R.A. 117), F-66650 Banyuls-sur-Mer, France.*

A. C. BROWN, *Department of Zoology, University of Cape Town, Rondebosch 7700, South Africa.*

E. D. HOUDE, *Chesapeake Biological Laboratory, University of Maryland, Solomons, Maryland, 20688, USA.*

J. M. E. STENTON-DOZEY, *Department of Zoology, University of Cape Town, Rondesbosch, 7700, South Africa.*

J. THÉODORIDÈS, *Laboratoire d'Évolution des Êtres Organisés, Université P & M Curie, Paris, France.*

E. R. TRUMAN, *Department of Zoology, University of Cape Town, Rondebosch 7700, South Africa.*

CONTENTS

Predation on Eggs and Larvae of Marine Fishes and the Recruitment Problem

K. M. BAILEY AND E. D. HOUDE

Recent Studies on Spawning, Embryonic Development, and Hatching in the Cephalopoda

S. V. BOLETZKY

vii

Parasitology of Marine Zooplankton

J. Théodoridès

Sandy-Beach Bivalves and Gastropods: A Comparison between *Donax serra* and *Bullia digitalis*

A. C. BROWN, J. M. E. STENTON-DOZEY AND E. R. TRUEMAN

Predation on Eggs and Larvae of Marine Fishes and the Recruitment Problem

K. M. Bailey and E. D. Houde

Northwest and Alaska Fisheries Center, National Marine Fisheries Service, 7600 Sand Point Way NE Seattle, Washington 98115 and Chesapeake Biological Laboratory, University of Maryland, Solomons, Maryland 20688

I. Introduction

The regulation of abundance in fish populations is a subject of outstanding debate in marine biology, and variable recruitment of new individuals to a population is recognized as one of the most important processes. Controversial questions center on learning at what stage relative recruitment levels are established and determining the principal causes of recruitment varia-

ADVANCES IN MARINE BIOLOGY,
VOLUME 25 ISBN 0–12–026125–1

bility. The objective of this review is to discuss predation as a factor that may regulate levels of recruitment or generate variations in it. We discuss predation on fish eggs and larvae as an ecological process and review what is known of its importance in the dynamics of fish populations.

Among Johan Hjort's (1914) enduring contributions to understanding population dynamics of marine fishes was his recognition that fluctuations in catches of commercially important species were due to changes in population abundance and not simply due to migrations of the species across large areas. He recognized that changes in abundance resulted largely from variable survival of new year classes before they were recruited to the fishery. He hypothesized that two mechanisms might operate in each life stage to cause year class strengths to fluctuate.

Hjort (1914) proposed that (1) larval transport away from favorable nursery areas and (2) mass starvation of larvae at the time that feeding must be initiated were potential causes of extreme variations in year-class strength. Recent theories have elaborated on Hjort's ideas of feeding conditions and focused on the distribution of larval food in the water column (Lasker, 1975) or the match-mismatch of spawning by fish and production of larval food (Cushing, 1972, 1975a). Transport mechanisms, oceanographic retention mechanisms, and the timing of spawning also have been implicated as being important in the recent literature (Parrish et al., 1981; Sinclair and Tremblay, 1984). Other theories suggest variable environmental conditions, where mechanisms are not clearly understood (Shepherd et al., 1984; Koslow, 1984) and environment-species dominance interactions (Skud, 1982). Hjort's theories and most recent attempts to develop a unifying theory to explain recruitment variations generally ignore predation as an important factor.

Only rarely has predation been recognized in past conceptual models dealing with recruitment variations (e.g., Ware, 1975; Shepherd and Cushing, 1980; Sissenwine, 1984), but it is increasingly viewed as an important factor influencing egg and larval survival. Cushing (1974) suggested that predation is the main cause of mortality in the early life of marine fishes, and that the inverse relationship between mortality rate and age (or size; Fig. 1) is the result of a reduced number of potential predators on later stages. A meeting of experts in 1975 concluded that starvation and predation are the main agents of larval mortality (Hunter, 1976). In reviewing predation on fish eggs and larvae, Hunter (1981, 1984) proposed that predation is probably the major cause of mortality for many species. Because recruitment patterns generally are established in the early life of fishes, it is apparent that a unifying theory to explain variations in recruitment and regulation of numbers must involve predation.

To manage exploited fishes, it is important to understand the causes of changes in abundance and to know whether the observed changes are natural

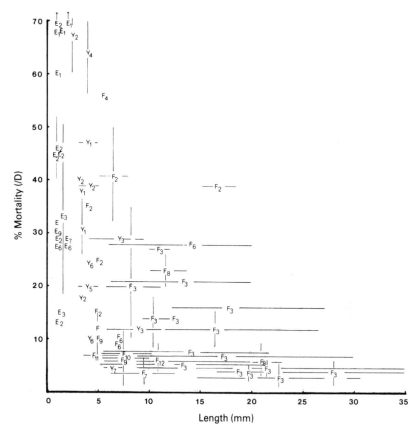

FIG. 1. Mortality rates of fish eggs (E), yolk sac larvae (Y) and feeding larvae (F) plotted against size. Numbers designate species. Values were obtained from the literature. Vertical bars are the ranges cited. Horizontal bars are size ranges over which the rates were determined. Species are (1) anchovy, (2) mackerel, (3) herring, (4) capelin, (5) flounder, (6) sardine, (7) plaice, (8) shad, (9) haddock, (10) saury, (11) cod, and (12) redfish.

or induced by man. Sissenwine (1984) noted that recruitment variability is the central problem of fishery science and a major source of uncertainty in management. Predation, including cannibalism, is especially important from a manager's perspective because it is often proposed as a cause of density-dependent regulation in stock-recruitment relationships (Ricker, 1954; Mac-Call, 1980; Rothschild, 1986). Multi-species interactions that focus on predator–prey relationships are of increasing interest, especially when fishing pressure on one species may influence the abundance of other species (Mercer, 1982; Gulland and Garcia, 1984).

The importance of invertebrate predation on fishes was apparently noted as early as the 18th century (Slabber, 1778; cited in Alvariño, 1985). Pioneering marine biologists of the early 20th century reported predation by invertebrates on fish eggs and larvae (Mayer, 1917; Joubin, 1924; Bigelow, 1926; Beebe and Tee-van, 1928: all cited in Alvariño, 1985). Bowman (1922) noted the occurrence of "spawny haddock", or haddock gorged with herring eggs. Lebour (1922, 1923, 1925) was among the first to experiment on predation of fish larvae. She reported that a diverse assemblage of zooplankton fed on marine fish larvae in the sea, and verified in laboratory experiments that these predators were capable of catching eggs and larvae and not just feeding in the net during collection.

Literature of the 1960s and 1970s is dominated by laboratory behavioral studies of predator–prey interactions and field observations of predator stomach contents. During this period, starvation as a cause of larval mortality was the primary focus of recruitment studies. Research on predation lagged behind because of its practical difficulty in both experimental and observational studies. In the late 1970s and early 1980s several international workshops and meetings recognized the importance of predation (see Sharp, 1980; SCOR, 1981; Bakun et al., 1982; Rothschild and Rooth, 1982; Sharp and Csirke, 1983; May, 1984) and stimulated renewed interest in it.

In this review we examine predation on fish eggs and larvae by invertebrate predators and fishes. Birds and mammals are largely excluded from our discussion. Our primary focus is on mid to high latitude marine fishes, but we have cited literature on tropical and freshwater fishes when appropriate. Initially we discuss the types of predators on fish eggs and larvae and assess how ecological and behavioral interactions influence the vulnerability of individuals and populations. We then review methods and discuss problems of studying the impact of predation on egg and larval survival in the sea. Finally, we discuss adaptations of fishes to avoid predation on early life stages, processes of starvation versus predation, and the role of predation in regulating fish recruitment.

II. Behavioral and Ecological Interactions of Fish Eggs and Larvae and their Predators

A. *The Diversity of Predators*

Fish eggs and larvae are eaten by many organisms in the ocean, ranging from the dinoflagellate, *Noctiluca scintillans* (Hattori, 1962) to marine birds (Palsson, 1984; Hunt and Butler, 1980). Greene (1985a, 1985b, 1986) attempted to understand mechanisms underlying patterns of prey selection in planktonic food webs, and thereby developed a conceptual framework for classifying predators by functional groups. These groups were based on the predator's methods of encountering, detecting and subduing prey. We have adapted his framework to our discussion of predators (Fig. 2). Later we use these functional groups to develop conceptual models of relative vulnerability of larvae to them.

According to Gerritsen and Strickler (1977), invertebrate and vertebrate predators have several fundamental differences, and this is the basis for the first functional division. Invertebrates select mostly for small prey due to limited prey-size capturing abilities. They rarely use vision to detect prey and they usually are only slightly larger than their prey. Vertebrates usually use vision to detect prey, they are generally much larger than planktonic prey, and they swim much faster than their prey.

1. Invertebrates

Ambush raptorial invertebrates include some gelatinous zooplankton and they are important consumers of fish larvae (Purcell, 1985). Fraser (1969) and others have described the feeding behavior of several ambush-type gelatinous zooplankton and estimated their potential to consume larval fishes. Cydippid ctenophores and all siphonophores, except *Physalia*, feed while drifting passively with their tentacles spread (Purcell, 1985). In general, they do not perceive prey before contact. *Pleurobrachia bachei*, for example, may at times swim actively, but when fishing for prey, its tentacles are extended and it drifts passively. Once a prey is entangled, tentacles are retracted and the prey is drawn into the mouth (Greene, 1985a). Some ambush-type cnidarians, such as siphonophores, may use lures to attract prey (Purcell, 1980).

Some small, cyclopoid copepods, such as *Corycaeus* spp. (Greene, 1986), in marine and freshwaters are raptorial ambush (or slow cruising) predators that use vision or mechano-reception to orient to potential prey. They can attach to and feed on fish larvae (Brewer *et al.*, 1984; Fabian, 1960; Lillelund, 1967). Smith and Kernehan (1981) reported *Cyclops bicuspidatus* to be a common predator on larvae of striped bass (*Morone saxatilis*) and white perch (*Morone americana*) in tidal fresh waters of the Chesapeake Bay. In the Laurentian Great Lakes, *Diacyclops thomasi* (= *Cyclops bicuspidatus*) and *Acanthocyclops vernalis* were significant predators on larvae of alewife (*Alosa pseudoharengus*), yellow perch (*Perca flavescens*) and other species (Hartig *et al.*, 1982; Hartig and Jude, 1984, 1988). Both copepods are believed to use mechanoreceptors to detect prey at a distance of 2–3 body lengths (Kerfoot, 1978) and to eat their prey bite by bite (Zaret, 1980). Nearly all predation by *D. thomasi* and *A. vernalis* on fish larvae occurred at night (Hartig and Jude,

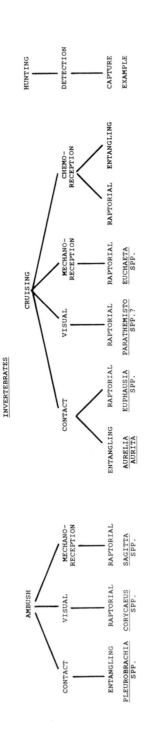

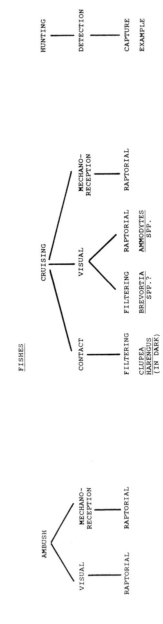

Fig. 2. Conceptual scheme showing types of predators of fish eggs and larvae, classified based on their hunting, detecting and capturing strategies.

1984). The marine cyclopoid copepod *Corycaeus anglicus* was found attached to 4.6% of white croaker (*Genyonemus lineatus*) larvae and 1.6% of northern anchovy (*Engraulis mordax*) larvae collected off the Californian coast (Brewer *et al.*, 1984). *Corycaeus* spp. reportedly uses vision to locate prey (Gophen and Harris, 1981; Landry *et al.*, 1985). However, most observed predation on white croaker larvae was at night. In contrast, anchovy larvae appeared to be equally vulnerable during the day. Brewer and coauthors sometimes found two or three *Corycaeus* attached to a larva and believed that many copepods dropped off after biting a larva. If that were the case, the incidence of predation by biting copepods might be higher than observed.

Chaetognaths are ambush-raptorial predators. They detect prey by mechanoreception and will respond to probes vibrating at specific frequencies (Feigenbaum and Reeve, 1977) as well as to tail beats of larval fishes (Kuhlmann, 1977). *Sagitta hispidia* moves in the water in a sink-swim up cycle (Feigenbaum and Reeve, 1977), but according to Kuhlmann (1977) *Sagitta setosa* and *Sagitta elegans* swim horizontally. After locating prey, chaetognaths strike quickly and engulf prey whole in their toothed jaws. Chaetognaths feed on fish larvae in the laboratory (Kuhlmann, 1977) and in the sea (Lebour, 1923; Stevenson, 1962; Brewer *et al.*, 1984). Because feeding rates are low and they prefer copepods, Kuhlmann (1977) and Hunter (1981) considered chaetognaths unlikely to be major predators on ichthyoplankton. But, the great abundance of chaetognaths in the sea makes them potentially important consumers of newly-hatched fish larvae (Brewer *et al.*, 1984).

Lobate ctenophores, cubomedusae, most scyphomedusae and many hydromedusae are actively swimming predators, which have no known ability to detect prey before contact (Purcell, 1985). The medusa *Aurelia aurita* is an important predator on fish larvae (Möller, 1980; 1984). *Aurelia* swims by propulsion of water from its pulsating bell-shaped body. Small *Aurelia* medusae capture fish larvae after they contact the exumbrellar surface, where larvae are stung by nematocysts and become stuck to mucus. Larger medusae may capture prey with tentacles (Fraser, 1969; Bailey and Batty, 1984). A recent, detailed description of *A. aurita* feeding on herring larvae is found in Heeger and Möller (1987). Lobate ctenophores, such as *Mnemiopsis leidyi*, use ciliary currents to draw prey such as larval fish into lobes where they are caught in mucus or secured by muscular action (Purcell, 1985).

Euphausiids are cruising-contact feeders (Mauchline, 1980) and may be important predators of fish larvae (Theilacker and Lasker, 1974; Bailey, 1984; Theilacker, 1988). Euphausiids swim with thoracic legs opening and closing while making many turns. Feeding currents are created by the activity of the pleopods and thoracic legs, which draw prey into the mouth parts. Mauchline (1980) reported no evidence that euphausiids hunt or stalk prey.

Chemosensory and visual detection of prey is poorly documented among the cruising invertebrate predators, but there is evidence that some gelatinous zooplankton, for example certain ctenophores (Swanberg, 1974) and crustaceans, such as the shrimp *Acetes sibouc australis* (Hamner and Hamner, 1977), are capable of chemosensory detection of prey. Other crustacean predators, such as euphausiids and amphipods, may use vision. Sheader and Evans (1975) reported visual predation by the amphipod *Parathemisto gaudichaudi*, but they suggested that there is also some chemical recognition of prey.

Yamashita *et al.* (1984) noted that *Parathemisto japonica* approaches Japanese sand eel (*Ammodytes personatus*) larvae before striking. The described behavior indicates visual orientation, but because *P. japonica* can feed at night as well as by day, there also may be chemical detection of prey. *P. japonica* and *P. gaudichaudi* approach larval prey with quick darting movements and grasp them by the midportion of the body with their periopods. The limbs are armed with spines to hold captured prey. *P. japonica* macerates its prey while hovering with its pleopods beating, either while sinking or resting on the bottom (Sheader and Evans, 1975; Yamashita *et al.*, 1984).

Westernhagen (1976) suggested that chemical detection of fish larvae may occur in the amphipod *Hyperoche medusarum*, but that actual attacks result from random encounters. In a later paper Westernhagen *et al.* (1979) attributed detection of prey to mechanoreception. Most studies on visual versus chemical detection of prey are based on cursory observations and should be verified in filmed experiments, as in Hamner and Hamner (1977).

Other amphipods known to feed on fish larvae include *Calliopius* (Bailey and Stehr, 1986) and *Gammarus* (Logachev and Mordvinov, 1979).

Some copepods are cruising raptorial predators. Predatory behavior of copepods in the genus *Euchaeta* (Yen, 1982; Bailey and Yen, 1983; Greene, 1985a; Yen, 1987) is especially well-documented, including observations of predation on marine fish larvae. A host of other copepod species that are potential predators on fish larvae have been identified by Lebour (1925), Lillelund and Lasker (1971), Corten (1983), and Turner *et al.* (1985). Most copepod predators detect prey by mechanoreception (Lillelund and Lasker, 1971; Zaret, 1980; Yen, 1983; Greene, 1985a). In at least one instance the authors claimed that prey are not detected, but are encountered randomly (Turner *et al.*, 1985). Copepods that detect prey from the fluid deformations caused by prey swimming have elaborate mechanoreceptor systems comprised of sensory hairs on their antennae (Yen, 1982). Although both cyclopoid and calanoid copepods use mechanoreception to detect prey, their hunting strategies differ. Based on Zaret's (1980) description, calanoid copepods swim above the wake of prey, dive and seize prey with their large maxillipeds, whereas cyclopoids follow the prey's wake and attack in a burst, stabbing it with daggerlike mandibles.

2. Fishes

Filter-feeding fishes such as juvenile or adult engraulids and clupeids (Hunter and Kimbrell, 1980; Brownell, 1985; Alheit, 1987) are among the best-documented predators. Filter-feeding fishes may continuously filter water with their mouths open, or pump filter by gulping and expelling water through their gills. Menhaden (*Brevoortia tyrannus*) swim with their mouths open and gill covers flared. Water entering the mouth is strained through fine apertures between the gill rakers before exiting through the branchial arches (Durbin and Durbin, 1975). Hunter (1981) and Colin (1976) have described the filter-feeding behavior of northern anchovy and the Indian mackerel (*Rastrelliger kanagurta*), respectively, on pelagic fish eggs.

Some fishes, for example menhaden, are obligate filter feeders (Durbin and Durbin, 1975), but many pelagic fishes, such as Atlantic herring (*Clupea harengus*), northern anchovy and Pacific mackerel (*Scomber japonicus*), may switch from filter feeding to raptorial feeding. The switch of feeding mode depends on prey size and density (Gibson and Ezzi, 1985; O'Connell and Zweifel, 1972). Facultative planktivores may feed by filtering or biting in daytime. However, at night prey must be encountered by random search and then filtered. Biting utilizes vision and it is restricted to daytime or moonlit nights in herring (Batty et al., 1986).

Planktivorous, raptorial fishes bite or suck prey into their mouths by expanding the buccal cavity and expelling water through the gills. As in filter-feeding fishes, small prey-like eggs and larvae are retained on the gill rakers. Feeding on fish eggs and larvae by raptorial fishes has been documented in the laboratory (Christensen, 1983; Brownell, 1985; Dowd, 1986; Margulies, 1986; Pepin et al., 1987) and in the field (Grave, 1981; see further references in Shapiro et al., 1988). In general, there are two types of attack strategies for raptorial fishes (Hunter, 1984). The lungers start an attack from an S-shape. They strike at high speed from short range, and seldom pursue missed prey. Pursuers start an attack from a C-shape. They strike at slower speed from a shorter distance and may chase missed prey. Raptorial planktivorous fishes can be effective predators. Seghers (1974) reported that the freshwater *Coregonus clupeaformis* can nip at rates up to 2400 times per hour.

Many fishes, including some gadids, are facultative planktivores. Although such fish may be inefficient feeders on small particles due to their large mouths (Liem, 1984), they could at times take substantial numbers of eggs and larvae, as Margulies (1986) found for juvenile bluegill (*Lepomis macrochirus*) preying on larvae of white perch.

Demersal fishes can be important predators on demersal eggs that are highly aggregated; for example, winter flounder (*Pseudopleuronectes americanus*) feeds heavily on eggs of capelin (*Mallotus villosus*) (Frank and Leggett, 1984). Because capelin eggs are buried, flounder must detect them without vision, possibly using chemoreception. Other demersal fishes that feed on eggs include saithe (*Pollachius virens*), cod (*Gadus morhua*) and haddock (*Melanogrammus aeglefinus*) (Dragesund and Nakken, 1973; Johanessen, 1980).

B. *Components of Predator–Prey Interactions*

Predation is viewed as a sequence of events, each of which can be analyzed and quantified (O'Brien, 1979). A conceptual model (Fig. 3) helps to structure our discussion of the components of predator–prey interactions with respect to fish eggs and larvae. An *encounter* is recorded when the prey is in the volume surrounding the predator in which it can be detected. Encounter rate is a function of population densities and swimming speeds (Gerritsen and Strickler, 1977; Gerritsen, 1980). *Detection* is a function of the "visibility" of the prey and the reaction distance of the predator. The latter depends on the sensory system used for detection. An encounter may or may not result in *pursuit, strike* and *capture*. Then the cycle may begin anew, depending on the satiation state of the predator. According to O'Brien

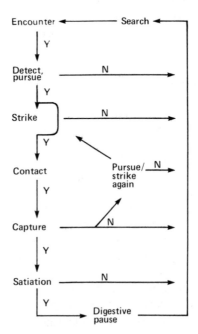

FIG. 3. Conceptual model of the predation process (adapted from Riessen *et al.*, 1984; Blaxter, 1986). N = no; Y = yes.

(1979), the importance of each component in the cycle varies with predator, prey and circumstance; behavioral and environmental factors contribute to the likelihood of success at each stage.

1. Encounter

Searching for and locating prey is probably the single most important phase of the predation cycle for organisms that are much larger than their prey, such as planktivorous fishes (O'Brien, 1979) feeding on eggs and larvae. Encounter rates depend on predator feeding mode, prey density, swimming speeds of predator and prey, size of prey and encounter radius of the predator (Gerritsen and Strickler, 1977; Gerritsen, 1980). Temperature can influence encounter rates through its effect on swimming speeds.

Zaret (1980) stated that for size-dependent predators (zooplankton by his definition), prey density is the most important feature influencing encounter frequency, especially when predator swimming speeds are low. Many laboratory studies have demonstrated the effect of larval fish density on predation rates by invertebrates (Theilacker and Lasker, 1974; Westernhagen and Rosenthal, 1976; Bailey and Yen, 1983; Bailey and Batty, 1983; de Lafontaine and Leggett, 1988; Monteleone and Duguay, 1988).

Densities of pelagic eggs usually are low but under certain circumstances can be quite high. For example, Hunter and Kimbrell (1980) reported mean density of northern anchovy eggs to be $32/m^3$ but patches of eggs with densities as high as $31\,000/m^3$ were observed. Densities of pelagic larvae are much lower, although on occasion they can form high density patches of newly-hatched larvae exceeding $100/m^3$ (e.g. Hewitt, 1981; Frank and Leggett, 1982). Patches of newly-hatched Pacific herring larvae, concentrated in the surface layer, have been sampled at densities of 750–2750/l (J. Purcell, University of Maryland, Horn Point, Maryland). High densities of demersal eggs (exceeding $10^6/m^2$ for Pacific herring eggs), whose dispersal rate is low, can contribute to high rates of predation loss. Predation rates by fishes on aggregated demersal eggs, e.g. capelin (Frank and Leggett, 1984) and Atlantic herring (Dragesund and Nakken, 1973; Johannessen, 1980) can be extremely high and have been recorded up to $20\,000$ eggs/predator/day (Johannessen, 1980). McGurk (1986) contended that the high mortality rates generally observed for eggs and larvae is due to predation, and is induced by the patchy nature of their distribution.

Cruising speeds of predators and those of prey are important. When predator and prey have different mean speeds, the faster of the two has the greater effect on encounter rates (Gerritsen and Strickler, 1977). In theory, slow-moving predators (e.g. chaetognaths) and immobile prey (e.g. fish eggs) have little opportunity for an encounter. This could explain why chaetognaths feed on young fish larvae, but not on eggs (Kuhlmann, 1977). Alternatively, because fish eggs are motionless relative to the surrounding water, they may not be detectable by the mechanoreceptors of chaetognaths. All other factors being equal, slow-moving and ambush predators should have highest encounter rates with fast-moving prey. For example, siphonophores are ambush predators that appear to capture larger, more active prey in higher proportion than such prey occur in the environment. Purcell (1981) attributed this "selectivity" to more frequent encounters. In a study comparing predatory behavior of freshwater predators on various prey species, Cooper et al. (1985) verified that for agile prey, both encounter rates and capture success were the most important determinants of predation rates. For non-evasive prey, encounter rates were most important. And, the apparent selection for mobile prey by ambush predators was due to higher encounter rates. In the case of fast-cruising predators, the prey's swimming speed had little effect on encounter rates, and predators selected for the less evasive prey.

Many cruising invertebrate predators on larval fish search at speeds greater than those of the newly-hatched larvae. Most fish larvae swim at average speeds of 1–2 body lengths/s or 3–10 mm/s for larvae 3–5 mm in length (Blaxter, 1986; Fig. 4). Predatory copepods can cruise at 1–4 body lengths/s, or 5–20 mm/s for a copepod 5 mm in length (Lillelund and Lasker, 1971; Greene, 1985a), and cruising speeds of amphipods are 10–60 mm/s (Westernhagen et al., 1979). Medusae of A. aurita with bell diameters of 5–25 mm swim at average speeds of 6–16 mm/s (Bailey and Batty, 1983). Most predatory fishes cruise at 1 body length/s or more, which is considerably faster than invertebrate predators. For example sand eel search for prey while swimming at 11 cm/s (Christensen, 1983), herring at 8.4 cm/s and sprat at 7.2 cm/s (Fuiman and Gamble, 1988).

Many predators change their swimming speeds after encountering prey. Sand eels (Ammodytes marinus) increased their swimming speeds fourfold after larval Atlantic herring were encountered as food (Christensen, 1983). Increased speeds also were recorded for A. aurita medusae after capturing larval herring (Bailey and Batty, 1983). Such behaviors increase encounter rates after the predator has located a patch of prey.

Changes in the swimming directions or in the numbers of turns have been recorded for predators foraging in a patch of prey. Area-concentrated searching is well documented for predators in general and can serve to keep them within a patch of prey (Curio, 1976). For example, A. aurita swam in a circling pattern after capturing larval herring (Bailey and Batty, 1983). Colin (1976) and Hunter (1981) noted that Indian mackerel and northern anchovy, respectively, swam in tight circles after encountering patches of fish eggs

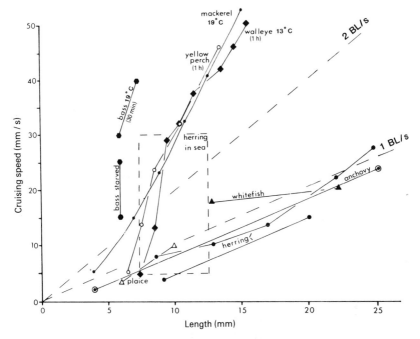

FIG. 4. Cruising speeds related to lengths of larval largemouth bass, yellow perch and walleye, Pacific herring, lake whitefish, Atlantic herring, plaice, northern anchovy, and Pacific mackerel. Values for bass, perch and walleye are from flume experiments for the duration indicated. Lines representing 1 and 2 body lengths (BL) per second are also drawn (from Blaxter, 1986).

upon which they fed. Batty *et al.* (1986) observed similar behavior in Atlantic herring filter-feeding on zooplankton.

Swimming directions can influence encounter frequency (Gerritsen, 1980). Laboratory experiments demonstrate that herring larvae, for example, do not swim in random directions, but tend to swim up and down in darkness and horizontally in the light (Batty, 1987). In theory, if predator and prey are swimming parallel to one another, then there is no chance for an encounter. Prey therefore can minimize encounters by this tactic; conversely, cruising predators can maximize encounters by swimming at right angles to the dominant direction of prey.

2. Prey detection

Invertebrate predators detect their prey using several different sensory systems. These include mechanoreception and, to a lesser extent, vision and chemoreception. Other invertebrates (i.e. cnidarians) apparently cannot detect prey before contact. Little is known about detection of fish eggs and larvae by invertebrate predators using visual or chemosensory systems. Detection of prey by mechanoreceptors is based on larval swimming characteristics.

Larvae display many different means of swimming, including: serpentine swimming (herring; Batty, 1984), high frequency tail beating that is nearly continuous or that may be intermittent (different developmental stages of northern anchovy; Hunter, 1972), and body undulations with pectoral fin movement (plaice; Batty, 1981). Whether these swimming modes confer an advantage to larvae is unknown. Butler and Pickett (1988) suggested that the beat and glide behavior of larval anchovy reduced vulnerability to predation by adult anchovy compared with the continuous swimming of sardine (*Sardinops sagax*). However, the advantage was related to escape behavior rather than detection.

Activity of larvae is another factor that may enhance the performance of invertebrate mechanoreceptors. Larvae display a wide range of activity levels that vary with species and stage of development (Blaxter, 1986). Many species are active 100% of the time (e.g. mackerels), sending out continuous signals to mechanoreceptive predators, while other species are less active (e.g. anchovy), reducing their "visibility" (Theilacker and Dorsey, 1980). Ontogenetic changes in activity also occur.

The detection of prey by predators using mechanoreception depends on the strength of the disturbance created by prey, the distance between the disturbance and predator, sensitivity of the predator's receptors, and the relative noise (such as turbulence) in the system (Gerritsen and Strickler, 1977). A moving predator creates a pressure shear around its body. This disturbance combines to reduce the sensitivity of its receptors by increasing the level of noise (Gerritsen and Strickler, 1977). Such disturbances increase with increasing speed, resulting in a reduction of the potential detection radius as the predator swims faster. Consequently, most invertebrate predators that use mechanoreception to detect prey, e.g. chaetognaths and the predatory copepod *Euchaeta*, must swim quite slowly while searching. Invertebrates that detect prey by visual or chemosensory mechanisms, or contact, can increase encounters by swimming fast (Gerritsen, 1980) probably without affecting detection ability.

The reaction distance of fishes to prey is greater than that of mechanoreceptive invertebrates. For example, sand eels react to larval herring at a distance of about 1 body length (Christensen, 1983) and cannibalistic anchovy respond to larval prey at 30–40 cm (Folkvord and Hunter, 1986). The freshwater fish, *Lepomis gibbosus* reacts to prey 3 mm in length at about 50 cm (Confer and Blades, 1975). In contrast, chaetognaths react to vibrating probes at distances no greater than 1–3 mm (Feigenbaum and Reeve, 1977) and the carnivorous larva of the freshwater midge, *Chaoborus americanius* reacts to *Daphnia pulex* at 1–3 mm distance (Riessen *et al.*, 1984). Little is known about the reaction distance of other mechanoreceptive predators, especially when feeding on fish larvae.

The detection of prey by visually-feeding fishes is determined by the apparent size of prey (Brooks and Dodson, 1965; Eggers, 1977). Apparent size is dependent on absolute size and proximity of prey (O'Brien *et al.*, 1976), pigmentation (Zaret and Kerfoot, 1975) and motion (Lindstrom, 1955; Kislalioglu and Gibson, 1976).

The effect of visibility on egg and larval susceptibility to fish predators is largely untested, but some studies support the importance of contrast in predator–prey interactions. For example, Brownell (1985) found that Cape anchovy (*Engraulis capensis*) eggs and yolk-sac larvae were less vulnerable to cannibalism due to their transparency than were older, eyed larvae. Folkvord and Hunter (1986) attributed relatively low rates of cannibalism by northern anchovy on yolk-sac larvae to transparency and lack of pigmentation. Dowd (1986) reported a similar effect when bay anchovy preyed upon eggs, yolk-sac larvae and older larvae of sea bream. With respect to larval size, Folkvord and Hunter (1986) reported that large larvae are attacked by fish predators from a greater distance than small larvae. Prey movement stimulates attacks by sand eels on Atlantic herring larvae (Christensen, 1983).

Behaviors of feeding fishes are varied. Searching strategies range from that of ambush predators, such as clownfish (*Amphiprion percula*) (Webb, 1981) to swim-searching of obligate planktivores such as blueback herring (*Alosa aestivalis*) to swim–pause–search behavior of facultative planktivores such as bluegill (Janssen, 1982) or white crappie (*Pomoxis annularis*) (O'Brien *et al.*, 1986). In darkness, filter-feeding fishes such as Atlantic herring may not detect prey before contact, but they nevertheless change behavior in the presence of patches of prey, swimming faster and swimming in tight circles (Batty *et al.*, 1986). This behavior also was noted for northern anchovy in the presence of prey extracts, implicating chemical recognition of prey (Hunter and Dorr, 1982).

Prey detection by visually-feeding fishes depends on prey contrast and brightness (Confer and Blades, 1975). Thus, turbidity, time of day and depth will influence prey detection. Other factors that influence predator selectivity are hunger level (Ivlev, 1961), the relative abundance of alternative prey, and image-searching behavior (Tinbergen, 1960; Ware, 1971). Differential evasion by prey may affect apparent selection and choice of prey may be influenced by prior experience. Fishes may correlate visual appearance with evasive capabilities of prey, leading to predator selectivity (Vinyard, 1980).

3. Capture and escape

Susceptibility (also known as capture success, the percentage of encounters resulting in ingestion; Greene, 1986), is determined by a prey's ability to detect and respond to the predator's presence or attack.

Fish eggs cannot actively escape attacks. Fancett (1988) attributed strong positive selection for fish eggs in their diet by scyphomedusae to this lack of escape ability. However, eggs have passive defense mechanisms that reduce their susceptibility to capture by small invertebrates. Eggs are large relative to the food of most planktonic invertebrates and have a resilient chorion that makes them difficult for small invertebrate predators to grasp and rupture (Bailey and Stehr, 1986; Bailey and Yen, 1983; Turner et al., 1985). The ultrastructures of egg cuticles are diverse and often complex. Some of these features may lend protection from predators, although Robertson (1981) thought that they were more likely to be adaptations to optimize ascent rates from depths of spawning. Of the many small crustacean invertebrates tested in the laboratory, only amphipods with large grasping appendages were capable of high predation rates on fish eggs (Bailey and Yen, 1983; Bailey and Stehr, 1986).

Larvae may successfully evade capture if they detect the predator. Larvae are equipped with several sensory systems including visual, mechanorecep-tive, auditory and tactile systems. However, these systems are not all functional throughout posthatching development (Blaxter, 1988).

Immediately after hatching, most marine fish larvae have unpigmented and non-functional eyes (Blaxter, 1986). At first feeding, the retinas of most larvae are comprised entirely of cones (Blaxter, 1986). As larvae develop further and visual ability improves, rods appear. They are important in perception of movement and therefore predator avoidance.

The apparent looming threshold (ALT) measures the visual response of larvae to approaching predators (Dill, 1974a; Webb, 1981). This is the rate of change of the angle subtended by an approaching object measured at the subject's eye. According to Webb, the looming effect becomes greater with increasing predator speed, closeness and size. The ALT decreases with increasing larval length due to improved acuity or maturation of neural pathways. Webb (1981) associated development of the ALT with escape behavior of northern anchovy larvae; in his experiments, the percentage of northern anchovy larvae attempting escape from attacks by clownfish increased from 9% for 2.9 mm larvae to 80% for 12.2 mm larvae (Webb, 1981; Fig. 5). Folkvord and Hunter (1986) obtained similar results for northern anchovy larvae responding to attacks by adult anchovy (Fig. 6) and Pacific mackerel (Scomber japonicus).

Young fishes may learn to respond visually to predators based on

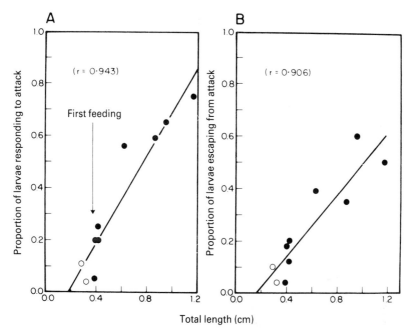

FIG. 5. Relationships between (A) the proportion of northern anchovy larvae responding to an attack by clown fish and (B) the proportion escaping an attack, both as functions of the total length. Circles are for pre-feeding larvae (from Webb, 1981).

experience, and thus increase their reactive distance (Dill, 1974b). Predators also may learn to attack successfully and capture larval fishes. Dowd (1986) believed that the sigmoid functional response of juvenile bay anchovy preying upon sea bream larvae reflected learning experience on the part of the predator at moderate to high densities of prey.

Neuromast organs probably detect water disturbances in the environment (Blaxter, 1986). Larvae can avoid predators that feed by suction (raptorial fishes), ciliary currents (*Mnemiopsis*), or those preceded by a bow wave (*Aurelia*, filter-feeding fishes) if they detect the hydromechanical disturbance and swim out of the flow field. However, newly-hatched herring larvae show no apparent response to water movement from gentle pumping of water toward them (Blaxter and Batty, 1985), even though neuromast organs are present (Blaxter *et al.*, 1983). Attempting to capture larvae with a pipette presumably should elicit escape behavior if neuromast organs can generate such a response. Success in capturing larvae with a pipette was high (80–100%) for larval yolk-sac stages of cod, flounder (*Platichthys flesus*), plaice (*Pleuronectes platessa*), herring and turbot (*Scophthalmus maximus*), but decreased as development advanced in these species (Bailey, 1984), possibly

Predator – *Engraulis mordax*

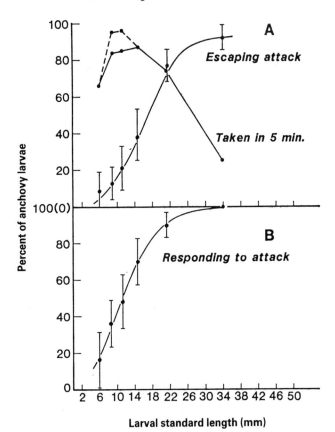

FIG. 6. Escape and response of northern anchovy larvae to adult anchovy as a function of larval length. (A) Percentage of northern anchovy larvae escaping attack. (B) Percentage of northern anchovy larvae that responded to the attack of an adult northern anchovy (from Folkvord and Hunter, 1986).

due to neuromast organ proliferation (e.g. Blaxter *et al.*, 1983). Yin and Blaxter (1987) found that 70–80% of herring and cod responded to the sucking action of a pipette. They were capable of a directional response away from the stimulus only after they developed to the feeding stage. Avoidance of pipette capture by these species was significantly correlated with their ability to avoid being eaten by the medusa *A. aurita* and the copepod *Euchaeta norvegica* (Bailey, 1984).

Young herring larvae, 10–12 mm in length, also showed no escape response to acoustic stimuli (100 Hz), although such behavior was observed for larvae longer than 22 mm (Blaxter and Batty, 1985). Herring larvae apparently can respond to acoustic stimuli only after the otic bulla contains gas. By contrast, zebrafish (*Brachydanio rerio*) larvae responded to acoustic stimuli soon after hatching (Kimmel *et al.*, 1980).

Young larvae may respond to tactile stimulation and to contact with predators. The percentage of herring larvae responding to contact with an inert probe increased with size (Blaxter and Batty, 1985). Yolk-sac larvae of cod, flounder, herring and turbot were mostly unresponsive (responding to 10–16% of contacts) compared with older larvae (responding to 12–68%) (Bailey, 1984). Yin and Blaxter (1987) obtained similar results. Differences among species appeared to be related to developmental stage. For example, yolk-sac plaice, which are relatively well-developed at hatching, responded to 81% of contacts.

The response to tactile stimuli involves the neuromast organs, or other touch or pain reception systems (Blaxter and Batty, 1985). The percentage of larvae responding to contact with *A. aurita* medusae, which presumably sting and induce pain, increased with larval size (Bailey and Batty, 1984). Small larvae, such as yolk-sac stages of cod and flounder, usually showed no obvious escape response to contact with medusae, but appeared to be paralyzed. This contrasted with larger herring larvae which responded to 97% of contacts with medusae. Reactions by herring larvae to stings of medusae involve violent contortions and "backing-off" (Blaxter and Batty, 1985), behaviors that are absent after contact with inert probes.

Fuiman (1986) noted that successful evasion by zebrafish larvae of a predator is a function of timing (response latency and looming threshold), magnitude (distance and velocity) and direction of the escape attempt. The interval between perception of stimuli and detectable response in young larvae, or the latency of the escape response, takes less than 0.1 s (Eaton *et al.*, 1977). Reaction times alone for zebrafish larvae are in the order of tens of milliseconds (Eaton and Hackett, 1984). For comparison, the reported time for expansion of the buccal cavity of a predator fish, largemouth bass (*Micropterus salmoides*), is about 0.04 s (Nyberg, 1971). Despite its importance, little is known about latency of escape response for species other than zebrafish. Webb (1981) observed that larval northern anchovy responded too late to avoid capture in 24–30% of encounters with clownfish.

The larval escape response is typically a fast–startle response, comprised of a C-shaped contortion and rapid acceleration (Eaton and DiDomenico, 1986; Blaxter and Batty, 1985). Startle responses are mediated by Mauthner cells or other brain neurons (Eaton *et al.*, 1977; Blaxter and Batty, 1985; Blaxter, 1986). Zottoli and Horne (1983) demonstrated a connection between the lateral line and Mauthner cells in adult goldfish. In the zebrafish the fast–startle response can be elicited in embryos by tactile or vibrational stimulation as early as 44 h after fertilization (Eaton and DiDomenico, 1986).

Burst swimming may be effective to escape predators that pursue prey. Burst swimming speed is a function of length (Fig. 7) developmental stage, and feeding condition (Yin and Blaxter, 1987). Acceleration of prey may influence the decision of predators to abort an attack or to chase prey (Webb, 1986). Predators apparently learn to recognize prey characteristics such as acceleration ability that serve to reduce capture success rate.

Several environmental factors may influence burst swimming speeds, the most obvious being temperature and water viscosity. Zebrafish larvae accelerated faster, travelled further per unit of time, and attained greater maximum velocity at higher temperatures (Fuiman, 1986).

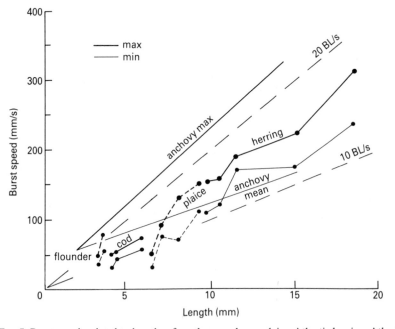

Fig. 7. Burst speeds related to lengths of northern anchovy, plaice, Atlantic herring, Atlantic cod, and European flounder. Maximum speeds are for a very short period after a flight response, mean speeds for a longer period. Lines representing 10 and 20 body lengths (BL) per second are also drawn (from Blaxter, 1986).

Generally, capture success by predators declines with increasing larval size due to the developing ability of larvae to respond to predators. Purcell *et al.* (1987) report that 90% of 8–9 mm yolk-sac Pacific herring larvae were captured after contact with tentacles of the hydromedusa *Aequorea victoria*, but only 37% of 13 mm and 13% of 19 mm larvae were captured. Bailey and Batty (1984) reported that 50% of contacts of early yolk-sac Atlantic herring larvae (9–10 mm) with 35 mm *A. aurita* medusae resulted in captures, where-as no contacts with 18 mm larvae were successful. That study also demon-strated clear inverse relationships between capture success and larval devel-opmental stage for flounder, cod and plaice. Starved herring larvae became increasingly susceptible to capture.

Heeger and Möller (1987) showed that herring contacting parts of *Aurelia aurita* were caught with different efficiencies; tentacles caught 94% of larvae contacted, whereas exumbrellar and subumbrellar surfaces caught 76 and 14% of larvae. In another study reporting capture success, Yamashita *et al.* (1984) reported that in 20 attacks on 1–3-day-old larval sand eel *Ammodytes personatus*, by the amphipod *Parathemisto japonica*, all were successful.

For fish predators, Brownell (1985) reported that 100% of cape anchovy eggs attacked by 29–35 mm anchovy juveniles were successfully captured. Successful attacks decreased with larval development, e.g. 92% of attacks on uneyed yolk-sac larvae were successful, and success declined to 60% on 10 mm larvae. Larvae longer than 17 mm were not captured. Folkvord and Hunter (1986) obtained similar results for northern anchovy and Pacific mackerel feeding on anchovy larvae, as did Butler and Pickett (1988) for northern anchovy feeding on sardine larvae. Dowd (1986) noted a steep decline in the proportion of sea bream larvae captured by juvenile bay anchovy predators as larvae grew from 2.2 to 4.5 mm. In this case, predation rate declined by 61% per mm of growth. Margulies (1986) also noted a steep decline in predation rate by juvenile bluegill on growing white perch larvae. All white perch larvae 3.5 to 6.0 mm long were captured by the predator, but susceptibility to capture declined subsequently by approximately 21% per mm for white perch 6 to 14 mm in length.

Folkvord and Hunter (1986) noted the effect of predator size and attack speeds on susceptibility of larval northern anchovy. For several predator species, larval escape ability always increased with larval size. Folkvord and Hunter observed that the smaller the predator, the faster larval escape ability improves with increasing size (Fig. 8). Anchovy larvae longer than 20 mm responded more frequently to northern anchovy attacks than to Pacific mackerel attacks because mackerel attacked at higher speeds. At higher attack speeds, larvae are less responsive because there is less time to react. Generally, larvae responded less frequently to large predators than to small ones because attack speed increased with predator size. Folkvord and Hunter further suggested that, in general, little or no predation occurs when larvae are longer than 50% of the length of their fish predators.

Larvae that are attacked by predators but not captured, may die if wounded. Lillelund and Lasker (1971) noted that yolk-sac northern anchovy larvae that had been wounded by predatory copepods never recovered. Turner *et al.* (1985) reached a similar conclusion for menhaden *Brevoortia*

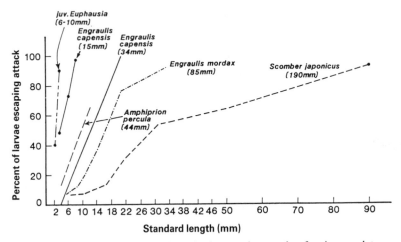

Fig. 8. Percentage of larval and juvenile anchovies escaping attacks of various predators as a function of length. Data for *Engraulis capensis* feeding on larval *E. capensis* are from Brownell (1985); juvenile *Euphausia* fed *E. mordax* from Theilacker and Lasker (1974); *Amphiprion percula* fed *E. mordax* from Webb (1981); and others are from Folkvord and Hunter (1986). Numbers indicate length (mm) of the various predators (from Folkvord and Hunter, 1986).

tyrannus larvae wounded by copepods. Westernhagen and Rosenthal (1976) reported that yolk-sac Pacific herring larvae that were grasped by the amphipod *Hyperoche medusarum* never survived after being released. Bailey (unpublished data) tested the recovery ability of yolk-sac herring, plaice, cod and flounder larvae to encounters with the medusa *A. aurita*. Herring and plaice larvae that had been stung by medusae survived for 24 h at least as well as controls that were handled similarly but which had not encountered medusae (herring: 100% survival for stung larvae versus 96% survival for controls; plaice: 93% survival for stung larvae and controls). Cod and flounder larvae were more vulnerable to stings (cod: 67% survival for stung larvae versus 80% survival for controls; flounder: 60% survival for stung larvae versus 77% survival for controls). Heeger and Möller (1987) reported that repeated contact of herring larvae with medusae resulted in death.

Hickey (1979; 1982) experimentally wounded herring larvae and found that some yolk-sac and first-feeding herring larvae survived amputation of up to 2 mm of the tail and incisions of 0.3 mm in the body. Skin removal of up to 1–3% of the total body surface area was also tolerated. Hickey noted that it is difficult to relate these findings to the sea. The wounds that were tolerated were surprisingly large compared with those that might be inflicted by small predators. The author concluded that newly-hatched herring larvae are most vulnerable to wounds, but that even these should survive some damage inflicted by invertebrate predators.

C. *Relative Vulnerability of Eggs and Larvae*

Vulnerability of individual prey to predators is a product of encounter rate and susceptibility (Greene, 1986). Greene (1986) and Zaret (1980) have illustrated theoretical changes in vulnerability curves for predator types, and shown effects of altering the predator–prey size ratio. We have adapted these curves to review vulnerability of eggs and larvae to predator types (Fig. 9).

Vulnerability to invertebrate ambush predators feeding on larvae of increasing size is often observed to be a bell-shaped function (Fig. 9(a)). Zaret (1980) believed that these bell-shaped predation curves were characteristic of prey size-dependent invertebrate predators, such as ambush invertebrates and predatory copepods. For example, vulnerability of eggs and early yolk-sac larvae to chaetognaths is low because they detect prey by mechanoreception and these prey stages are relatively inactive. Vulnerability of larvae to chaetognaths increases to a maximum as larvae develop and their swimming activity increases. Thereafter, susceptibility declines as larval escape ability improves (Kuhlmann, 1977). Finally, large larvae are invulnerable to chaetognaths because of the limited gape of the jaw of the predator (Pearre, 1980). These same arguments were advanced by Bailey and Yen (1983) to explain laboratory observations of changes in predation rates by the slow-cruising copepod *Euchaeta elongata* on larvae of Pacific hake (*Merluccius productus*).

In the case of cruising invertebrates that do not use mechanoreception exclusively to detect prey, such as some amphipods, euphausiids and medusae, increases in prey size and activity result in small increases in encounter rates because the predator's speed greatly exceeds that of the prey. Susceptibility and vulnerability to these predators are quite high because early stages of fish larvae have poorly developed predator detection and escape abilities. Declining vulnerability of progressively larger larvae results from improved detection of predators and better ability to escape (Fig. 9(b)). As examples, declining vulnerability curves for predators feeding on larvae were observed for the euphausiid *Thysanoessa raschi*, and the medusae *Aurelia aurita* and *Aequorea victoria* feeding on larvae in tanks (Bailey, 1984; Bailey and Batty, 1984; Purcell *et al.*, 1987), as well as for medusae in field studies (Möller, 1980; Purcell, 1986). In another example, Yamashita *et al.* (1984) reported that older larvae of Japanese sand eel were less vulnerable to capture by the amphipod *Parathemisto japonica* than were 1–3-day-old larvae. In mesocosm experiments similar to naturally occurring conditions, de Lafontaine and Leggett (1988) found that predation by the medusa *Staurophora mertensi* on capelin larvae did not follow the above trend. They reported a significant positive relationship between larval mortality and larval size, suggesting that the larger larvae (late yolk-sac and early feeding

24 K. M. BAILEY AND E. D. HOUDE

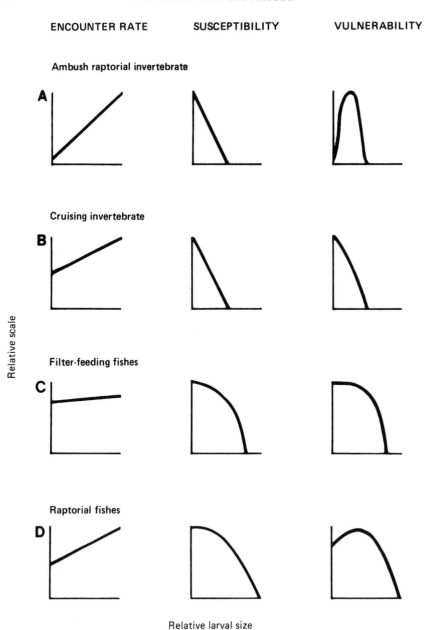

FIG. 9. Conceptual models showing relative encounter rates, larval susceptibility, and vulnerability of larvae to different predator types.

stages) were more susceptible to jellyfish predators compared with early yolk-sac stages. They could not distinguish whether increased predation resulted from higher encounter rates or reduced escape ability.

There are few data on vulnerability of eggs and larvae to filter-feeding fishes. Increases in prey swimming speed have little effect on encounter rates. Assuming that encounters are random and that gill-raker spacing for most filter-feeders is less than 0.5 mm, then most pelagic eggs and early-stage larvae are vulnerable (Fig. 9(c)). Leong and O'Connell (1969) reported that northern anchovy filter fed on *Artemia* nauplii that were 0.6 mm long and anchovy could retain particles less than 0.1 mm. Durbin and Durbin (1975) indicated that apertures between gill rakers of menhaden were irregularly spaced but probably averaged less than 0.1 mm. There is probably little difference in prey detection or selection due to prey size within the available size range of pelagic fish eggs and yolk-sac larvae. Because the approach of a fish predator is much faster than the escape velocity of young larvae, predator detection and escape response may not be significant until larvae are older. Prey size eventually will constrain most filter-feeding fishes because they have relatively small mouths. Durbin and Durbin (1975) reported that maximum prey size of menhaden is between 1.5 and 10 mm.

Raptorial-feeding fishes that visually detect prey are analogous to the gape-limited predators of Zaret (1980). In this case, encounter rate increases with prey size because of increased prey visibility. Susceptibility of fish larvae initially decreases slowly as larvae grow, but then declines rapidly when prey size approaches the limitations of the predator's mouth gape (Fig. 9(d)). Trends of increasing vulnerability as prey size increases have been shown by many investigators (see Zaret, 1980), and most recently by Pepin et al. (1987) for Atlantic mackerel feeding on various larvae. Brownell (1985) reported that juvenile cape anchovy will cannibalize the largest anchovy larvae possible. Folkvord and Hunter (1986) presented prey vulnerability curves for northern anchovy cannibalizing larval anchovy similar to that in Fig. 9(d). Yolk-sac anchovy larvae were less vulnerable because of low visibility; maximum predation rate was attained on first-feeding larvae before improved escape abilities or predator gape limitation caused a decrease in vulnerability.

A general summary of size-dependent vulnerability to different types of predators is presented in Table 1. Exceptions to this scheme exist and species-specific differences in behavior and activity affect vulnerability. Yet, there are at least three notable results in Table 1:

(1) the relative invulnerability of eggs to some predator types.
(2) the high vulnerability of yolk-sac larvae to all predator types and

TABLE 1. VULNERABILITY OF EGGS AND LARVAL SIZE CLASSES TO DIFFERENT PREDATOR TYPES

| | | | | Predator Type | | | | |
| | | | | Cruising-Raptorial | | | Fishes | |
Prey	Ambush Gelatinous	Cruising Gelatinous	Ambush Raptorial	Large (chemo/visual)	Random Encounter	Small (mechano)	Raptorial	Filtering
Eggs	—	***(*)¹	—	**	***(—)²	—	*	***
Yolk-sac larvae	**	***	**	***	***	**	**	***
Larvae 3–5 mm	**	**	**	**	*	*	***	***
Larvae 5–10 mm	*	*	*	*	*	—	***	**
Larvae 10–20 mm	—	—	—	—	—	—	***	*
Larvae >20 mm	—	—	—	—	—	—	**	—

¹Depends on nematocyst type.
²Depends on predator size and capture method.

(3) the probable great importance of fishes as predators on early life stages.

Filter-feeding fishes can be effective predators on the smallest larvae while raptorial fishes can be important throughout the larval and juvenile stages.

D. *Population Vulnerability*

Vulnerability of egg and larval populations to predator populations is a function of individual vulnerability of prey to specific predators, overlapping of predators and prey in space and in time, the ingestion capacity of predators and the response of predators to prey density. Ingestion capacity of predator populations depends on predator abundance, predator gut capacity and the rate of digestion of prey. These factors may be modulated by environmental conditions such as temperature, turbidity, turbulence, light, vertical distributions, and the abundance of alternative prey.

1. Spatial and temporal overlap

The spatial overlap of predators and prey affects vulnerability of larval populations. Frank and Leggett (1985) have pointed out that even if a predator is potentially dangerous to larvae, such larvae are often spatially separated, either by occupying different water masses or by different vertical positions in the water column.

Many predators of eggs and larvae migrate vertically; this behavior may change depending on season or food availability. In Puget Sound, Washington, the copepod *Euchaeta elongata* lives near the seabed in daytime and migrates vertically into the surface layer at night to feed. But, during late winter and spring a part of the *Euchaeta* population may feed near the seabed by day and night (Yen, 1983). Based on the food bolus characteristics and its black coloration, *Euchaeta* probably feeds on fish larvae in that period. Chaetognaths and euphausiids also may migrate vertically. Many cnidarian species migrate vertically, although some do not and can be distributed throughout the water column (Mills, 1982).

Many invertebrate predators, as well as fish eggs and larvae, are present in the water column for only a part of the year. Thus, temporal overlap is critical. Spawning of Pacific hake *Merluccius productus* in Puget Sound commences prior to the spring phytoplankton bloom (Fig. 10), but ends before peak abundances of most predator populations (the predators shown may not necessarily be predators on hake eggs and larvae). Hake eggs are spawned at 100–150 m and remain near the seabed through egg and yolk-sac

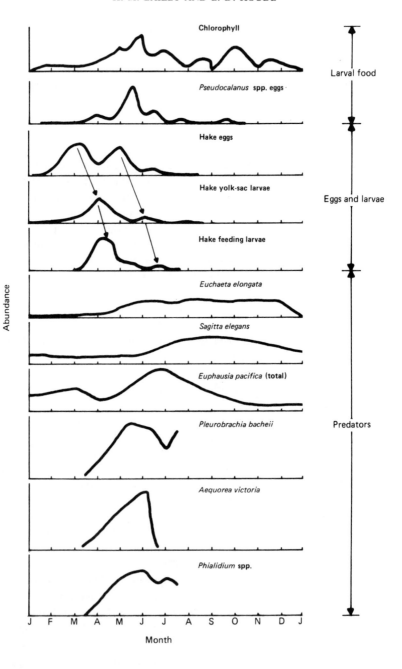

stages until larvae begin to feed (Bailey, 1982). This distribution could be an adaptation to avoid predation of eggs by filter-feeding fishes, such as Pacific herring, in the near-surface water of Puget Sound. Although eggs probably are not vulnerable to many invertebrate predators, except amphipods and certain medusae (Bailey and Yen, 1983), yolk-sac larvae are vulnerable to many invertebrates that occur throughout the water column. Optimally, spawning should occur just before the spring phytoplankton bloom to assure abundant larval food, e.g. copepod nauplii (Runge, 1981), but well before predator populations become abundant. In 1973, hake eggs that were spawned late in spring, compared with those spawned earlier, had poor survival to the feeding stages (Bailey and Yen, 1983), possibly because yolk-sac larvae in June were decimated by many predators that co-occurred at that time (Fig. 10).

Another example in which the temporal and spatial overlap of larvae and predators may be important is that of plaice and flounder larvae in the Dutch Wadden Sea (Veer et al., 1983; Veer, 1985). Most plaice larvae are transported into the Wadden Sea before blooms of gelatinous zooplankton, whereas flounder larvae enter later and co-occur with high abundances of medusae and ctenophores. Veer proposed that differences in timing of immigration contribute to the relatively high predation mortality of flounder larvae compared with plaice.

Even though potential predators and fish larvae overlap in space and time, subtle differences in timing of reproduction can result in changes in the predator–prey size ratio. Purcell (pers. comm.) noted that herring spawn in British Columbia waters when abundances of ctenophores and medusae are relatively low, and moreover, when the available predators are small. Early in the spawning season, most hydromedusae and chaetognaths are too small to capture herring larvae. By April, when both size of predators and their numbers have increased, predators can consume most herring larvae that are produced (Purcell, 1986). de Lafontaine and Leggett (1988) hypothesized that predation by macroinvertebrates regulates survival of larval capelin (*Mallotus villosus*) and that annual variation in larval survival is due to variability in the seasonal emergence of larvae and the growth of jellyfish.

Variations in timing of both the spring bloom and peak abundances of predators can influence survival of larvae. Annual differences in the abun-

---

←— FIG. 10. Generalized scheme for population dynamics of hake eggs and larvae, their food, and predators in Puget Sound, Washington. Data for chlorophyll *a* and *Pseudocalanus* eggs (nauplii would be similar) are from Ohman (1983); hake *Merluccius productus* eggs and larvae from Bailey and Yen (1983); *Euchaeta elongata* from Yen (1982); *Sagitta elegans* and *Euphausia pacifica* from Ohman (1986); and *Pleurobrachia bachei, Aequorea victoria,* and *Phialidium* spp. in the Strait of Georgia, British Columbia from Larsen (1986).

dances of medusae and ctenophores of 2–100 fold are common (Möller, 1984; Veer, 1985; Frank, 1986; Purcell, 1986). Fraser (1970) reported differences in timing of peak abundance of *Pleurobrachia pileus* of three months between consecutive years in Scottish waters. Frank (1986) reported order-of-magnitude variability in abundances of *P. pileus* on the Scotian shelf in 1983 and 1984. He suggested that the year-class failure of haddock in 1983 may have been precipitated by abundant *P. pileus*, which consumed potential food of haddock larvae. In this case neither Frank (1986) nor Koslow *et al.* (1985) thought that ctenophores caused massive haddock mortality by predation but thought that indirect effects via competitive feeding may have been important.

2. Consumption capacity of predator populations

In addition to larval vulnerability (encounter rate, detection, capture success), the ingestion capacity of predator populations depends on predator abundance, prey handling times, gut capacity, satiation, digestion rate, and rate of growth in response to larval abundance. Environmental factors such as temperature also can affect feeding capacity of a predator population.

Handling time, the interval between capture and ingestion of prey, has been reported for invertebrates feeding on eggs and larvae in only a few instances. Yamashita *et al.* (1984) reported that the amphipod *P. japonica* requires 10–40 s to position sand eel larvae in front of its mouth. Purcell (1981) observed that ingestion of a larva by the siphonophore *Rhizophysa eysenhardti* required about 8 min. She suggested that the feeding rate of tentaculate predators may be limited by both prey handling time and space within the gastrozooids. Turner *et al.* (1985) observed that the handling time of the copepod *Anomalocera ornata* feeding on menhaden larvae was 2–3 min. Handling time by fish feeding on planktonic organisms much smaller than themselves, such as fish eggs and larvae, is short (Eggers, 1977) and probably insignificant unless each predation event is followed by a pause before the next feeding sequence begins.

Digestion times of predators feeding on eggs and larvae are influenced by many factors including temperature, activity, gut fullness, and prey size. In the case of invertebrates feeding on eggs and larvae, these factors have not been examined thoroughly. A summary (Table 2) indicates that, in general, digestion times for invertebrates feeding on fish larvae range from 3 to 8 h. Some invertebrate predators such as the ctenophore *Mnemiopsis* may have considerably faster rates (Monteleone and Duguay, 1988). Digestion rates for fishes feeding on larvae are very fast. Christensen (1983) reported that herring larvae are identifiable in guts of sand eels for only 15–30 min. Hunter and Kimbrell (1980) found that adult northern anchovy digest anchovy larvae in less than 30 min and eggs in about 3 h.

TABLE 2. DIGESTION TIMES OF INVERTEBRATES FEEDING ON FISH EGGS AND LARVAE

Predator	Prey	Time to digest	Reference
Medusae			
Rathkea octopunctata	Misc. larvae	14–19.5 h	Plotnikova, 1961
Sarsia tubulosa	Misc. larvae	3–3.5 h	(cited in Purcell,
Tiaropsis multicirrata	Misc. larvae	18 h	1985)
Aequorea victoria	Herring larvae	2–4 h	Purcell, 1985
Aurelia aurita	Herring larvae	3.5–9.5 h	Möller, 1980
A. aurita	Herring larvae	4–6 h	Bailey and Batty, 1984
Ctenophores			
Mnemiopsis leidyi	Anchovy eggs	10–15 min	Monteleone and Duguay, 1988
M. leidyi	7 day anchovy larvae	20–80 min	Monteleone and Duguay, 1988
Siphonophores			
Rhizophysa eysenhardti	Misc. larvae	3–7 h	Purcell, 1981
Physalia physalis	Misc. larvae	7 h	Purcell, 1984
Forskalia spp.	Misc. larvae	4–7 h	Purcell, 1983
Chaetognath			
Sagitta spp.	Misc. larvae	0.5–8 h	Kuhlmann, 1977
Crustaceans			
Parathemisto gaudichaudi	Sand eel larvae	2 day	Sheader and Evans, 1975
Euchaeta norvegica	Cod larvae	4–6 h	Yen, pers. comm.
Fish			
Ammodytes marinus	Herring larvae	15–30 min	Christensen, 1983
Engraulis mordax	Anchovy larvae	30 min	Hunter and Kimbrell, 1980
E. mordax	Anchovy eggs	3–12 h	

3. Predator response to prey density

The numerical response of predators describes how the number (or size) of predators increases or declines in response to prey density (Solomon, 1949). Over a short time (from a few hours to days), predator populations may respond to high density patches of prey by aggregating locally through immigration. For example, Frank and Leggett (1984) noted that predatory winter flounder respond rapidly to capelin spawning by aggregating near-shore where capelin eggs are abundant. Palsson (1984) observed that high densities of Pacific herring eggs attract bird flocks, possibly resulting in compensatory mortality of eggs. A possible numerical response by predatory *Crangon* shrimp on newly-settled plaice, as observed by Veer and Bergman (1987), may play a significant role in the complex process of population regulation (Fig. 11(a)).

Numerical response of predator populations to prey density over a longer time period (days to years) can occur if specific prey are an important fraction of the total diet. Capelin eggs formed a significant portion (59% by weight) of winter flounder diet for only 40 days, but their contribution of ingested energy to annual growth of flounder was substantial, about 23% (Frank and Leggett, 1984). Feeding by flounder on capelin eggs could contribute to its growth, reproductive success, and population abundance, a numerical response that could have long-term consequences in regulating capelin egg abundance.

The growth response (in biomass and numbers) of gelatinous zooplankton to prey abundance can be dramatic. For example, Veer and Oorthuysen (1985) reported that *Aurelia* medusae may increase their bell diameter by four times in 30 days. This can be important because medusa size has a major effect on predation rate on larvae (Bailey and Batty, 1984; de Lafontaine and Leggett, 1987). Medusae such as *Phialidium*, *Aglantha* and *Aequorea*, all

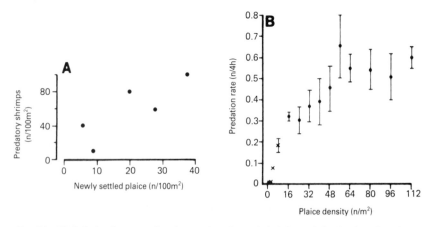

FIG. 11. (A) Relation between abundance of newly-settled plaice and the density of predatory shrimps on tidal flats in the western Wadden Sea. (B) Functional response of the predation by 10 shrimps in relation to plaice density for plaice of 10–15 mm length (redrawn from Veer *et al.*, 1987).

larval fish predators, can increase their population biomass by 10–100 times in 30 days (Larson, 1986) while ctenophores can double their weight each day (Greene, 1985a).

Gelatinous zooplankton can achieve extremely high densities in the sea. Purcell (1986) reported average and maximum densities of *Aequorea victoria* of 1.5 and 17 individuals/m³, respectively. Peak seasonal densities of *P. pileus* in the Dutch Wadden Sea average about 18 individuals/m³ (Veer et al., 1983). Yasuda (1970; cited in Möller, 1980) reported densities of *Aurelia aurita* up to 596 medusae/m³.

Larvae could be an important source of energy for many gelatinous zooplankton, contributing to their growth and reproduction. *R. eysenhardti* fed exclusively on fish larvae during a two-month bloom in the Gulf of California (Purcell, 1981). Fish larvae comprised 70–90% of the gut contents of *Physalia physalis* in the Gulf of Mexico (Purcell, 1984), and contributed up to 48% of the food of *A. victoria* in British Columbia waters (Purcell, pers. comm.).

The functional response of predators describes the number of prey eaten per predator as a function of prey density (Holling, 1959). Subtle changes in the functional response curve can cause marked changes in the shape of the mortality curve of the prey (Murdoch, 1973; Peterman and Gatto, 1978; see also Hildén, 1988).

Most vertebrate (Brownell, 1985; Hourston et al., 1981; Dowd, 1986) and invertebrate (Lillelund and Lasker, 1971; Theilacker and Lasker, 1974; Bailey and Yen, 1983; Bailey and Batty, 1984; Monteleone and Duguay, 1988; Fancett and Jenkins, 1988) predators which have been tested, increase their predation rate with increasing egg and larval densities. Functional responses may be asymptotic due to satiation or handling time constraints. A typical, asymptotic functional response curve relating predation rate by *Crangon* to newly-settled plaice density (Fig. 11(b)) indicated that shrimps essentially fed at maximal rates when plaice densities exceeded 48/m² under the experimental conditions provided by Veer and Bergman (1987). It is unclear what form of functional response curve many predators have at naturally occurring prey densities. Peterman and Gatto (1978) believed that most vertebrate predators have a sigmoid type curve. Dowd (1986) fitted sigmoid functional response curves to predation data by juvenile bay anchovy on sea bream larvae. Linear responses have been shown for cnidarians at low, naturally occurring prey densities (Fancett and Jenkins, 1988; de Lafontaine and Leggett, 1988; Monteleone and Duguay, 1988).

Ingestion capacities of gelatinous zooplankton and fishes in laboratory studies are extremely high compared with those of crustacean predators (Table 3). Also laboratory rates of predation are high compared with observations from field studies. There are several reasons for this: laboratory

TABLE 3. MAXIMUM PREDATION RATES ON FISH EGGS AND LARVAE IN LABORATORY STUDIES, AND AVERAGE RATES IN FIELD STUDIES

Predator	Prey	Prey eaten/t	Prey Density/l	Reference
LABORATORY STUDIES				
Invertebrates				
Aurelia aurita	Herring larvae	12/h	3/l	Bailey and Batty, 1984
Mnemiopsis leidyi	Anchovy larvae	2/4 h	1/l	Monteleone and Duguay, 1988
Mnemiopsis leidyi	Anchovy larvae	24/4 h	10/l	Monteleone and Duguay, 1988
Sagitta setosa, S. elegans	Sprat, anchovy larvae	4/10 h	50/l	Kuhlmann, 1977
Anomalocera sp.	Menhaden larvae	10/day	30/l	Turner *et al.*, 1985
Euchaeta elongata	Hake larvae	2–6/day	5–10/l	Bailey and Yen, 1983
Euchaeta norvegica	Cod larvae	6–10/day	10/l	Yen, 1987
Labidocera jollae	Anchovy larvae	18/day	8.6/l	Lillelund and Lasker, 1971
Parathemisto japonica	Sand eel larvae	24/day	40/l	Yamashita *et al.*, 1984
Parathemisto gaudichaudi	Sand eel larvae	47/2 day	3/350 ml	Sheader and Evans, 1975
Hyperoche medusarum	Herring larvae	1/h	100/l	Westernhagen *et al.*, 1979
Thysanoessa raschi	Cod larvae	3/day	3/l	Bailey, 1984
Euphausia pacifica	Anchovy larvae	17/day	11–14/l	Theilacker and Lasker, 1974

Fishes

Engraulis mordax	Anchovy eggs	125/h	38/1	Hunter and Kimbrell, 1980
E. mordax	Anchovy eggs	17 000/day	10/1	Hunter and Dorr, 1982
Anchoa mitchilli	Seabream larvae	44.5/h	2/1	Dowd, 1986
Lepomis macrochirus	White perch larvae	10/15 min	0.3/1	Margulies, 1986
Clupea harengus	Herring larvae	14 000/day	—	Hourston *et al.*, 1981
Ammodytes marinus	Herring larvae	80/5 h	0.1/1	Christensen, 1983
A. marinus	Herring larvae	12/10 s	0.1/1	Christensen, 1983
Clupea harengus	Herring larvae	13.5/day	0.025/1	Fuiman and Gamble, 1988

FIELD STUDIES

Invertebrates

Chrysaora melanaster	Misc. larvae	239/8 h	—	Nair, 1954
Physalia physalis	Misc. larvae	120/day	—	Purcell, 1984
Rhizophysa eysenhardti	Misc. larvae	8.8/day	—	Purcell, 1981

Fishes

Engraulis ringens	Anchovy eggs	239/day	—	Alheit, 1987
E. mordax	Anchovy eggs	85.8/day	—	Hunter and Kimbrell, 1980
Sprattus sprattus	Herring larvae	177/day	0.025/1	Fuiman and Gamble, 1988
Ammodytes tobianus	Herring larvae	150/day	0.025/1	Fuiman and Gamble, 1988

prey densities usually are higher than those in the sea, experiments are usually of short duration (whereas field observations integrate feeding over many hours), and the walls of containers may facilitate trapping of prey. Laboratory predation rates cited in Table 3 should be considered near maximum potential ingestion rates.

We have found little solid evidence that satiation normally occurs for predators on eggs or larvae in the sea. Peterman and Gatto (1978) suggested that predators feeding on juvenile salmon operate at the low end of their functional response curves, and thus maximum predation rates were not being attained. Purcell (pers. comm.) noted that the number of fish larvae in the guts of *Aequorea victoria* was proportional to the number of larvae in the surrounding waters, indicating that medusae were not satiated at such densities of larvae. The linear response of cnidarians cited above indicates that no satiation is observed at reasonable prey densities; however, in other mesocosm experiments Frank and Leggett (1982) did find that ctenophores approached satiation at larval capelin densities of 400–1000/m³, which is within the range that newly-hatched capelin are reported to occur in the sea. Furthermore, Heeger and Möller (1987) reported from laboratory experiments that 20 mm diameter *Aurelia aurita* medusae became satiated with 8 to 15 herring larvae in their gut, and the frequency of bell contractions decreased. Higher numbers have been found in guts of medusae caught in field studies. Crustacean zooplankton, small cnidarians and chaetognaths with their small gut volumes and slow digestion rates possibly may satiate in high densities of larval prey.

4. Effects of alternative prey

Presence of alternative prey affects predation rates of some invertebrates feeding on fish larvae in the laboratory. Alternative prey lowered predation rates by copepods on eggs or larvae (Lillelund and Lasker, 1971; Bailey and Yen, 1983; Yen, 1987), chaetognaths (Kuhlmann, 1977), and amphipods (Sheader and Evans, 1975). But, *Artemia* nauplii did not influence predation rates by *Euphausia pacifica* on northern anchovy larvae (Theilacker and Lasker, 1974). For gelatinous zooplankton, Monteleone and Duguay (1988) found that the presence of alternative prey did not affect predation by the ctenophore *M. leidyi* on anchovy eggs, and de Lafontaine and Leggett (1988) found similar results for several species of medusae preying on capelin larvae. Few studies of prey selectivity exist; chaetognaths preferred copepods to fish larvae (Kuhlmann, 1977), but the amphipod, *Parathemisto gaudichaudi* selected fish larvae over copepods (Sheader and Evans, 1975).

For fish predators, Brownell (1985) found that in laboratory experiments alternative prey (copepods) decreased cannibalism rates by juvenile cape anchovy on larvae and noted that both satiation and abundance of alternative prey affected the cannibalism rate. Pepin *et al.* (1987) found that adult Atlantic mackerel preferred fish larvae to zooplankton and Christensen (1983) reported that sand eels virtually ignored copepods as prey in the presence of herring larvae. Dowd (1986) observed a strong preference by bay anchovy juveniles for two-day-old larval sea bream relative to copepod prey jeuntil the ratio of copepods:larvae exceeded 33:1, after which the predator strongly selected copepods. Margulies (1986) found five to tenfold reductions in predation rate by juvenile bluegills on white perch larvae when *Daphnia magna* was added to experiments. But, there was no selection by bluegill for either white perch larvae or *Daphnia*. Both prey were of approximately equal size and were consumed in proportion to their abundances.

Experiments testing effects of alternative prey clearly demonstrate its importance in moderating the potential impact of many predators on early life stages of fishes. Because laboratory experimental conditions differ greatly from conditions in the sea, laboratory results must be interpreted with caution to understand how a predator may affect egg or larval abundance when alternative foods are available.

Pepin (1987) modeled the influence of alternative prey abundance on pelagic fish predation of northern anchovy larvae. He reasoned that alternative prey abundance will affect the time required for a predator to obtain its maximum ration. From his model, Pepin concluded that larval mortality should be a function of predator abundance scaled to relative alternative prey abundance.

To our knowledge, the influence of alternative prey on predation rates on fish eggs and larvae in the sea has not been demonstrated. In theory, if prey density were high enough for the predator to make a choice, the rate of predation on one type of prey must decrease if it co-occurs with a less evasive prey (Vinyard, 1980), especially if the predator satiates quickly. For many invertebrates, where the radius of prey detection is small, fish egg or larval prey may not be abundant enough for active selection, but will be attacked and consumed at every encounter. Alternative prey may still influence rates of predation because of the limited gut capacity and slow digestion time of prey by small predators. For vertebrates, alternative prey abundances should influence the time to attain maximum ration (Pepin, 1987) and if prey are distributed in layers or patches, could influence foraging behavior.

III. Predation on Fish Eggs and Larvae in the Sea

A. Methods of Assessing Predation

Methods to assess predation mortality have been difficult to develop.

38 K. M. BAILEY AND E. D. HOUDE

Problems have included:

(1) identifying prey in predator guts; (2) quantifying prey consumed by each predator; (3) estimating abundances of predators; and (4) estimating numbers of eggs or larvae available as prey.

1. Identification of prey

Eggs, which generally have a chorion that retains integrity during digestion, may be identifiable as fish eggs in fish stomachs for 3–12 h (Hunter and Kimbrell, 1980; Daan et al., 1985). Larvae can be identified in fish stomachs, but because they have thin integuments and relatively little calcified tissue, digestion rates of small individuals by predacious fishes are very fast, perhaps less than 1 h (Hunter and Kimbrell, 1980; Daan et al., 1985). Larger larvae (10–20 mm) are generally more resistant to digestion. Small larvae that were eaten by trawl-caught fishes may be mostly digested during a trawl tow of 45 min or more, but larger larvae and juveniles may be undigested, potentially biasing estimates of size-specific predation rates.

Gelatinous zooplankton also ingest their prey whole, but most species apparently digest them slowly, making identification of eggs or larvae in their guts relatively easy compared to predacious fishes. Gastric evacuation times for most gelatinous zooplankton feeding on fish larvae range from 3–8 h (Purcell, 1981, 1984).

Identification of fish eggs and larvae in guts of invertebrate predators that macerate their prey is particularly difficult. Larval fish remains have been detected visually in guts of euphausiids, amphipods and copepods (Alvariño, 1985). Remains of larvae that have been identified microscopically include otoliths (Yen, 1987; Bailey, unpublished data), eye lenses (Purcell, 1984), flesh (Yamashita et al., 1985) and melanin pigments (Bailey and Yen, 1983; Yamashita et al., 1985; Yen, 1987). Gastric evacuation times for larval prey appear to be variable among macerating invertebrate predators, ranging from 0.5 to 12 h.

Immunoassay techniques show promise for analyzing gut contents of predators that macerate their prey. These techniques may allow prey identification at the species level (Feller et al., 1979; Waddy and Aiken, 1985; Walter et al., 1986). Theilacker et al. (1986) applied an enzyme-linked immunoassay to assess euphausiid predation on northern anchovy yolk-sac larvae. The method requires isolating an antigen, in this case yolk protein, and developing an antibody to it. The presence of yolk in the predator gut was assayed by adding antiserum (from rabbit blood) and measuring the amount of antibody–antigen complex. Drawbacks may include lack of specificity and difficulty in translating yolk quantity to larval numbers because the amount of yolk per egg or larva declines as development proceeds and the amount in predator guts decreases with time due to digestion.

Other problems of identifying and quantifying prey in predator stomachs include feeding by predators in the net and voiding of gut contents during collection or preservation. Nicol (1984) and Lancraft and Robison (1980) reported net feeding on assorted prey by euphausiids and pelagic fishes, respectively. Veer (1985) noted that 95% of individual ctenophores, P. pileus, voided their stomach contents after fixation in formalin.

2. Estimating consumption

Once predation upon fish eggs and larvae has been established, it is necessary to estimate the daily ingestion by predators and then to partition it into specific prey items. There has been a proliferation of recent literature on methods to estimate daily consumption and gastric evacuation rates (e.g. Eggers, 1977; Elliott and Persson, 1978; Jobling, 1981, 1986; Pennington, 1985; Olson and Mullin, 1986; Persson, 1986), many of which are applicable to estimate predation on fish eggs and larvae.

A direct method to determine daily ingestion is to quantify the amount of food in stomachs during a time period and account for gastric evacuation during the period. Several such methods are available, but virtually all investigators of predation on fish eggs and larvae have used Bajkov's (1935) equation,

$$I = 24W/t$$

where I = daily ingestion, W = average stomach content over the feeding period, t = the time for prey to be totally evacuated from the stomach, and 24 (hours) = the feeding period. It is assumed that there is no periodicity in feeding, that gastric evacuation rate is linear and constant, and that evacuation rate is unaffected by the amount and type of food consumed. Furthermore, stomach samples must be collected frequently and randomly throughout the feeding period. Olson and Boggs (1986) pointed out that if feeding for 24 h was assumed, ingestion rates would be overestimated by 50% if feeding ceased during a 12 h night, and stomach samples were only collected in the daytime.

Elliott and Persson (1978) demonstrated that Bajkov's equation may underestimate food consumption of fishes compared with often more realistic models using exponential rates of evacuation. They recommended the equation,

$$I = at[S - S_o \exp(-at)]/[1 - \exp(-at)]$$

where S_o = weight of stomach contents at time $t = 0$; S = weight of stomach contents; a = rate (/h) of gastric evacuation; t = h/d that a fish feeds. Elliott and Persson's equation has not been applied to predation upon fish eggs and larvae, but Dwyer *et al.* (1987) have used this model to calculate consumption of O-group juvenile walleye pollock (*Theragra chalcogramma*) by adult pollock. The Elliott and Persson's model generally requires rigorous sampling intervals of 1–3 h. Boisclair and Leggett (1988) compared the Elliott and Persson and the relatively simpler Eggers' model in an *in situ* experiment and found that the two models were comparable in magnitude and accuracy of daily ration estimates for adult yellow perch (*Perca flavescens*). The Eggers' model may be preferred because of less demanding sampling requirements.

Another model is that of Olson and Boggs (1986),

$$I = 24\sum_i (W_i/A_i)$$

where i = food type, W_i = weight of a food type i in stomachs, and A_i = the integral of the function that best fits gastric evacuation data for food type i. This model does not require an *a priori* assumption of an evacuation rate model. Evacuation rates of individual prey types are accounted for in the model by assuming that each prey type is evacuated independently, although experimental evidence for this is equivocal (Persson, 1984; Olson and Boggs, 1986).

Eggers (1977) and Pennington (1985) provided theoretical arguments justifying Bajkov's equation, but with an exponential rate of gastric evacuation rather than a linear rate. Hunter and Kimbrell (1980), who estimated egg cannibalism by northern anchovy, and Yamashita *et al.* (1985), who examined predation by the amphipod *P. japonica* upon sand eel larvae, used a modified form of Bajkov's equation to determine ingestion rates. Because ingestion rates are dependent on the model selected, and particularly upon evacuation rates, choice of a model is important. Persson (1986) advocated curvilinear models. Jobling (1986) recommended linear models when prey are "large, high energy particles" and curvilinear models for "small, low energy food particles". Fish eggs or larvae might be in either category, depending upon the type of predator. Juvenile or adult planktivorous fishes probably consume fish eggs or larvae as small particles while many invertebrates may process each egg or larva as a relatively large meal.

Daily ingestion requirements of predators, from which potential predation impact on egg/larval populations could be determined, can be estimated indirectly from an energetic model,

$$I = (E + G)/A$$

where, I = ingestion, E = energy required for metabolism, G = energy required for growth and reproduction, and $A =$ assimilation coefficient. Stomach content analysis on predators to determine the contribution of eggs or larvae to the ration can be applied, from which the number consumed can be estimated. Without modification, application of this model requires the assumption that all prey have equal gut residence times. Stomach contents can be adjusted for differential rates of gastric evacuation (Olson and Boggs, 1986). Based on estimates of predator growth, several investigators have calculated the number of fish larvae needed for the predators to meet energetic requirements. Fraser (1969) calculated that an individual *A. aurita* medusa might consume 400–500 fish larvae in six weeks. Arai and Hay (1982), Veer (1985) and Veer and Oorthuysen (1985) also used this approach to estimate potential consumption of fish larvae by gelatinous zooplankton. Johannessen (1980) used the energetic requirements of cod and haddock to estimate their potential consumption of herring eggs.

Elliott and Persson (1978) argued that direct estimates of food consumption based on stomach contents require fewer assumptions than estimates from theoretical energy budgets. Bioenergetic models may suffer from errors in parameter estimation (Bartell *et al.*, 1986), but they are very useful to estimate effects of temperature, body size and activity on consumption rates (Olson and Boggs, 1986). Stomach contents methods are more direct, but also more labor-intensive, and they are subject to sampling problems which can result in infrequent numbers of point estimates of feeding rate. The suitability of either approach depends on the nature of predators and prey. In the case of fishes feeding on eggs, the stomach content approach might be better. In cases where stomach contents are not easily identified or digestion rates are unattainable, the bioenergetic model may be better.

Once the ingestion rate of an individual predator has been determined, the estimated abundance of predators is required to calculate the number of eggs or larvae consumed. Examples can be found in Hunter and Kimbrell (1980), Yamashita *et al.* (1985) and Mehl and Westgard (1983). Data of this type are difficult to obtain, especially for patchily-distributed, evasive predators, such as pelagic fishes and euphausiids, and fragile organisms, such as siphonophores and ctenophores. More effective nets, improved acoustical assessment of pelagic nekton, and development of non-destructive methods to collect gelatinous zooplankton will improve estimates of predator abundance.

Quantifying the abundance of eggs or larvae is also difficult. A high proportion of small larvae may be extruded through 333 µm and 505 µm meshes that are commonly used on plankton nets (Lo, 1983), and an increasing proportion of large larvae become effective avoiders of nets (Smith, 1981). Abundance estimates of eggs or larvae often must be adjusted before either total mortality or predation mortality can be estimated. Murphy and Clutter (1972) compared abundance estimates of anchovy

Stolephorus purpureus larvae caught with a plankton purse seine to those collected in a 1 m ring net and found that the purse seine was 10 times more efficient at catching larvae greater than 5.5 mm length in daytime than the ring net. Those results were used by Yamashita *et al.* (1985) to correct abundance estimates of sand eel larvae caught with a 1 m net. Leak and Houde (1987) also used the Murphy and Clutter results to adjust for avoidance of bongo net samplers by bay anchovy larvae.

3. Indirect methods

Because predation, which may result from several predators, is difficult to assess directly, indirect methods have been attempted. One method assumes that total mortality and starvation mortality can be estimated and that predation mortality can be calculated as the difference between them, assuming that other sources of mortality are not significant. This approach was used by Hewitt *et al.* (1985) on larval jack mackerel (*Trachurus symmetricus*). Leak and Houde (1987) also applied a modification of this approach to determine the relative impact of starvation and presumed predation mortality on bay anchovy larvae. In this method the predators are not identified and two mortality coefficients (total and starvation) must be estimated, both of which may be inaccurate or imprecise. A desirable feature is that total predation mortality is assessed, not just that inflicted by one or two predator species.

Another indirect method to investigate predator–prey relationships is from inverse correlations between the distributions of predators and prey (Ali Kahn and Hempel, 1974; Alvariño, 1980, 1981; Möller, 1980, 1984; Veer, 1985; Houde *et al.*, 1986). Inverse relationships between the spatial distribution of northern anchovy larvae and that of potential predators, such as siphonophores and medusae have been reported (Alvariño, 1980, 1981). Northern anchovy larvae were relatively more abundant in tows dominated by potential food, such as small copepods and euphausiids, than in tows where predators were abundant. Alvariño (1980) noted that such relationships must benefit survival of anchovy larvae.

The interpretation of asynchronous oscillations in predator and prey populations as evidence for predator–prey interactions has been criticized (Frank and Leggett, 1985; Leggett, 1986). Inverse correlations between numbers of predators and prey do not necessarily indicate causality, because physical mechanisms or temporal asynchrony in reproduction can segregate predators and prey.

Laboratory ingestion rates, combined with field observations have been used to infer predator impacts (Theilacker and Lasker, 1974; Bailey and Yen, 1983; Palsson, 1984; Koslow *et al.*, 1985; Fancett and Jenkins, 1988;

Monteleone and Duguay, 1988). For example, Bailey and Yen (1983) assumed that individuals of the copepod *Euchaeta elongata* with melanin in their guts had eaten hake larvae. They calculated the number of larvae eaten per day from the density of copepods, the percent with melanin in their guts, and rates of ingestion determined in laboratory experiments. Bailey and Yen (1983) concluded that *E. elongata*, as well as other predators, potentially could consume all available hake larvae during the later part of the spawning season when predator populations were high. Theilacker and Lasker (1974) reached similar conclusions regarding the potential impact of *Euphausia pacifica* feeding on larval northern anchovy, as did Palsson (1984) for snails and amphipods feeding on the demersal eggs of Pacific herring. Koslow *et al.* (1985) used such methods to show that *Pleurobrachia pileus* did not cause significant mortality of haddock eggs, and Fancett and Jenkins (1988) found similar results for two species of scyphozoans feeding on fish eggs in Port Phillip Bay, Australia. Finally, Monteleone and Duguay (1988) calculated from average population densities in a coastal area from Cape Hatteras to Cape Cod, US, and predator clearance rates from tank studies, that the ctenophore, *Mnemiopsis leidyi* could consume significant numbers of pelagic fish eggs. Reports to date must be considered preliminary. They illustrate a potential way to determine the impact of specific predators on eggs and larvae if the laboratory-estimated predation rates are representative of rates occurring in the sea.

4. Predation in mesocosm experiments

Large enclosures of 1.0 to $100\,000\,m^3$, termed mesocosms, may be an effective tool to investigate predation on fish eggs and larvae under conditions intermediate between small laboratory tanks and the open sea. Mesocosms may be large tanks, plastic or mesh bags suspended in the sea, excavated basins, or closed embayments. The use and potential of mesocosms in marine, larval fish research have been reviewed by Øiestad (1982), Gamble (1985), and Houde (1985).

Major advantages of mesocosms over smaller laboratory containers include the ability to stock foods, larvae and predators at near natural densities in systems where predators and prey may behave more naturally than in smaller tanks. In particular, wall effects are reduced although not eliminated, while experimental manipulations remain feasible. Disadvantages are related to size, scale, diminished physical structure, and reduced turbulence. The lack of advection or limited opportunities for vertical migration or segregation from predators could influence predation rates.

Rates of predation by gelatinous plankton apparently are lower in mesocosms than in smaller laboratory tanks. Gamble *et al.* (1985) found that

A. aurita in 5 m³ mesocosms consumed herring larvae at a rate of 4–20 larvae/ day when larvae were present at 40/m³. These rates are high but the authors noted that they were 2.5 times lower than laboratory-derived rates for *A. aurita* in 5 litre containers stocked with 15 herring larvae/l. Recently, de Lafontaine and Leggett (1987) reported a clear relationship between meso- cosm volume and predation rate by *A. aurita* on capelin larvae. Predation rate declined by a factor of 10 when mesocosm volume was increased from 0.26 to 6.3 m³ (Fig. 12). The authors noted that most of the decline occurred when mesocosm volumes increased from 0.26 to 3.0 m³, and suggested that mesocosms need not exceed 3.0 m³ to estimate the predation potential of *Aurelia* on larvae in the sea. The decline in predation rates in large volume mesocosms indicate that refuges, either through spatial or behavioral isola- tion of prey from predator, probably exist on a relatively fine scale in the sea. Monteleone and Duguay (1988) found an effect of container volume on predation rates of the *Mnemiopsis leidyi* feeding on anchovy eggs, but they found higher clearance rates in larger containers, probably due to disruption of feeding activity by increased contact with walls of smaller containers. Mesocosms that vary in their volume to depth ratio, but which have similar

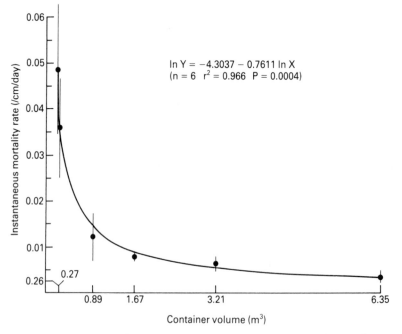

FIG. 12. Instantaneous mortality rate of capelin larvae due to predation by *Aurelia aurita* plotted against container volume (redrawn from de Lafontaine and Leggett, 1987).

surface area, like those used by de Lafontaine and Leggett (1987), can be important to define depth-mediated relationships between early life stages of fishes and their predators. Expanding such experiments to test predation in mesocosms of varying surface areas and depths also would provide important knowledge about the nature and scale of predator–prey interactions.

B. The Impact of Predators on Egg and Larval Populations—Specific Examples

Siphonophores are major predators of fish larvae (Purcell, 1981, 1984). Rates of predation on fish larvae based on Bajkov's formula for stomach content data indicated that each *Rhizophysa eysenhardti* in a small cove in the Gulf of California consumed 8.8 larvae/animal/day. Daily consumption was equal to 28.3% of the available fish larvae (Purcell, 1981). In another study, where larval fish comprised 70–90% of the diet of *Physalia physalis*, Purcell (1984) estimated that the population might consume 60% daily of the available fish larvae at a single site in the Gulf of Mexico.

Purcell (1986) examined predation by gelatinous zooplankton on herring larvae in coastal British Columbia, Canada and found that predator abundance was highly variable. There were especially large populations of predators in bays and inlets. The medusa *A. victoria* was an important predator, occurring at densities up to 17 animals/m³, and up to 80% of its diet was herring larvae. Yolk-sac and first-feeding larvae were most vulnerable to *Aequorea* predation. Purcell calculated that in one bay in 1985, the medusae consumed from 3 to 50% of available larvae hourly throughout an 8-day period.

Möller (1980, 1984) examined predation by the jellyfish, *A. aurita*, on Atlantic herring larvae. *Aurelia* occurred at densities up to 596 medusae/m³. He found as many as 68 larvae in the stomach of a 68 mm (bell diameter) medusa, and 10 larvae in a 12 mm medusa. Mean numbers of larvae per medusa ranged from 0.2 to 4.4. Most of the prey were yolk-sac larvae, 5–7 mm in length. Möller (1980) calculated that medusae consumed from 2–5% of the available herring larvae per day, and concluded that *Aurelia* was a major cause of larval herring mortality. Furthermore, inverse relationships between the abundances of herring larvae and medusae in both space and time were found that imply possible predator–prey interactions (Möller, 1984). Low numbers of herring larvae were found where

(1) the abundance of *Aurelia* was high,
(2) during days when *Aurelia* biomass was high (Fig. 13A), and
(3) in years when the average abundance of *Aurelia* was high (Fig. 13B).

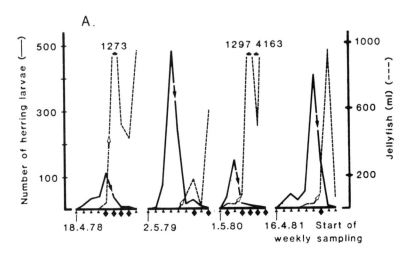

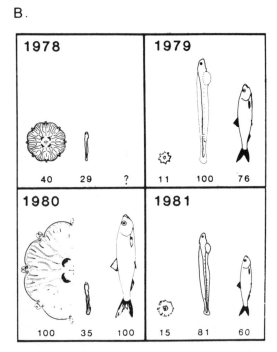

FIG. 13. (A) Average abundance of yolk-sac herring larvae (number /100 m³) and of *Aurelia aurita* (ml/100 m³) from Kiel Fjord sampled at weekly intervals. (B) Drawings symbolize the relative abundances of *Aurelia aurita*, yolk-sac herring larvae and adult spawning herring during four years in Kiel Fjord (from Möller, 1984).

Predation by *P. pileus* and *A. aurita* on plaice and flounder larvae in the Dutch Wadden Sea was investigated by Veer and Zijlstra (1982), Veer *et al.* (1983), and Veer (1985, 1986). It was concluded that for plaice larvae, predation by *P. pileus* was relatively unimportant, but predation on flounder larvae could be substantial. Flounder larvae are often 10 times more abundant than plaice, but after settlement plaice are about four times more abundant than flounder on tidal flats. Because plaice appear in the Wadden Sea in February, flounder in April and the gelatinous zooplankton blooms in May, the authors hypothesized that *P. pileus* develops too late to regulate plaice larvae numbers.

In support, only flounder larvae were found in the guts of ctenophores. Although the incidence of flounder larvae in preserved *P. pileus* guts was low (0.05%), laboratory experiments showed substantial regurgitation during fixation. Freshly caught *P. pileus* had a higher feeding incidence of 2.5%. The low percentages were due in part to the low ratio (1:100) of flounder to ctenophores in the sea. Given a digestion time of several hours and a density of *P. pileus* of 14 000/1000 m^3, they calculated that 140 larvae/1000 m^3 would be consumed in just a few hours. The average density of flounder larvae was only 215/1000 m^3. The calculations suggest that flounder larvae could be made extinct if new larvae were not transported into the Wadden Sea each day. Veer (1985) also noted a dramatic drop in abundance of flounder larvae with increasing biomass of coelenterates and concluded that the timing of larval immigration into the Wadden Sea may be important for survival.

Because herring eggs are demersal, they are not vulnerable to most pelagic predators. But, birds (Palsson, 1984), fishes (Johannessen, 1980) and benthic invertebrates (Palsson, 1984) can consume enormous numbers of herring eggs. Palsson monitored mortality rates of Pacific herring egg cohorts derived from discrete spawning events. He established study plots in spawning locales and subsampled by diving every 2–4 days. Daily egg-loss rates varied from 16.9 to 51.8%, and were positively correlated with initial egg densities. Experiments where large predators, such as diving ducks and gulls were excluded by nets from the study plots showed that in three of six study plots, large predators accounted for 20–50% of the daily egg-loss rate. Palsson concluded that egg-loss rates were mostly due to small invertebrate predators such as snails and caprellid amphipods. Their predation potential was determined from laboratory experiments as 5.6 and 8.0 eggs/individual/day. Estimates of *in situ* egg densities combined with potential egg-consumption rates showed that predation loss could have equalled or exceeded the initial cohort egg-density values. Earlier predator exclusion studies had demonstrated that bird and fish predators were responsible for 30–80% of herring egg loss (Outram, 1958; Steinfeld, 1972).

Predation on sand eel larvae by *Parathemisto japonica* varies seasonally and may be an important source of mortality (Yamashita *et al.*, 1985). Sand eel larvae comprised from 0 to 8% by weight of the *P. japonica* diet. The proportion of the larval sand eel population lost daily to predation by *P. japonica* was calculated to be 0.1% in January, 3.3% in February, 2.8% in March and 45.2% in April. Changes in predation pressure were attributed to changes in the size of sand eel larvae, abundance of *P. japonica* and degree of overlap of the two organisms. The high predation estimate in April may be exaggerated because the abundance of sand eel larvae was underestimated. In April many larvae were large enough to avoid the 1 m ring-net samplers but apparently were still vulnerable to amphipod predation.

Brewer *et al.* (1984) reported incidences of predation upon eggs and larvae by zooplankton. Overall, 5% of white croaker (*Genyonemus lineatus*) larvae and 2% of northern anchovy larvae had zooplankton predators attached to them. Three copepods (*Corycaeus anglicus*, *Labidocera trispinosa* and *Tortanus discaudatus*), one larval euphausiid (*Nyctiphanes simplex*), and one amphipod (*Monoculoides* sp.) were found attached to fish larvae. In one sample a 23% incidence of attached copepods was reported. Brewer *et al.* (1984) attempted to minimize biases related to net feeding or release of attached predators during collection, by conducting short bongo net tows and otter trawls. They reported the predation potential of copepods to be very high. One species, *C. anglicus*, was 10 times more abundant than all fish larvae. In some samples the three copepod species were 100 times more abundant than fish larvae. Because fish larvae occurred in only 7% of adult and juvenile demersal fish stomachs that they examined, Brewer *et al.* (1984) did not consider them to be important predators, but pelagic planktivorous fishes, which were not sampled, might have been more significant.

Cannibalism by the northern anchovy may be a significant source of egg mortality (Hunter and Kimbrell, 1980). 42% of adults sampled at night and 88% of those sampled in the day contained eggs. Larval fishes occurred in only 2% of stomachs. The daily egg production consumed was 17.2%. Because the natural mortality of northern anchovy eggs is about 53%/day, egg cannibalism caused 32% of the egg mortality. Hunter and Kimbrell also made plankton tows in front of and behind an anchovy school and found that eggs were 48% less abundant behind the school, presumably due to cannibalism. They concluded that patchiness of eggs and selective feeding may be important in regulating the impact of cannibalism on the population.

Cannibalism also may be significant in other pelagic fishes. Alheit (1987) reported that egg cannibalism accounted for 21.9% of the 68% per day egg mortality in Peruvian anchoveta (*Engraulis ringens*). He noted that sardine (*Sardinops sagax*) also were important predators on anchoveta eggs, but was unable to estimate their contribution to daily egg mortality. Cannibalism by 15–35 mm larvae of Cape anchovy on eggs and small larvae was significant in laboratory experiments and in a modelling exercise (Brownell, 1985, 1987) suggesting to the author that density-dependent regulation of recruitment via cannibalism might be possible. Valdés *et al.* (1987) in fact calculated from field data that cannibalism may account for 70% of the total egg mortality, and that it is density-dependent.

Hewitt *et al.* (1985) attempted to measure predation mortality on jack mackerel off the Californian coast as the difference between total and starvation mortality. Mortality rates of eggs and yolk-sac larvae were high, which the authors attributed to predation because starvation could not occur in the presence of adequate endogenous food. They estimated that mortality from fertilization to final yolk-sac resorption was from 99.5 to 99.9% (Fig. 14). The authors noted that estimated starvation mortality, determined by histological examination, exceeded the total mortality rate immediately after the onset of first-feeding. Thus, by the difference assumption, predation mortality dropped from 50 to 80% per day to zero, and then rose again to 20–30% per day a few days later. Small sample sizes and numerous assumptions required to make the calculations may have produced errors in estimates of total and starvation mortality. Hewitt *et al.* (1985) concluded that starvation of first-feeding jack mackerel larvae was signifi-

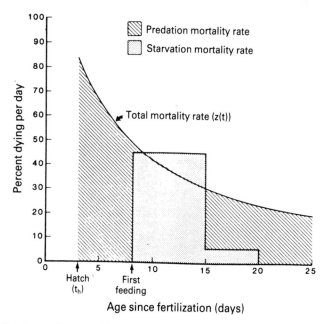

FIG. 14. Total mortality rate of larval jack mackerel *Trachurus symmetricus* as a function of age since fertilization. Mortality due to predation + mortality due to starvation = total mortality (from Hewitt *et al.*, 1985).

cant, but that predation was more important. Although there were problems in application, the approach is novel and promising.

Leak and Houde (1987) used a similar approach to partition total mortality of otolith-aged bay anchovy larvae into starvation and presumed predation components. Egg mortality averaged 85.6% per day but mortality of yolk-sac and older larvae was lower, averaging 26–36% per day. The starvation component, estimated from a laboratory model which predicted survival relative to food concentration, accounted for only one-third to one-half the total mortality in Biscayne Bay, suggesting that predation was the major cause of mortality during both the embryonic and larval stages.

Three possible reasons for the 1960s gadoid outburst in the Northeast Atlantic were proposed:

(1) relaxed competition from declining abundance of pelagic fishes,
(2) relaxed predation pressure due to declining pelagic fish predators on gadoid eggs and larvae, and
(3) changes in recruitment success and distributions due to climatic shifts (Cushing, 1980, 1984).

Results of studies on presence of gadoid eggs in pelagic fish stomachs were inconclusive (Daan, 1976; Garrod and Harding, 1981; Pommeranz, 1981). Daan et al. (1985) attempted to quantify egg predation rates by herring through extensive gut analyses. They calculated that of the initial number of plaice and cod eggs spawned, only 0.7–1.9% of plaice eggs and 0.04–0.19% of cod eggs were consumed by herring. Herring were not abundant during the years of their study, but they speculated that during peak herring abundance in the 1960s the maximum predation mortality would have been about 10% for plaice and only 1% for cod.

Among the best evidence that predation is the dominant source of mortality to eggs and larvae is that obtained in mesocosm research. High survival and growth rates of herring, capelin, cod and flounder larvae at food abundances similar to those in the sea have been obtained routinely in mesocosms when predators were absent (Houde, 1985). Øiestad (1985) summarized a decade of Norwegian basin experiments that clearly demonstrated the devastating effect of fish or invertebrate predators on enclosed populations of herring, plaice, capelin and cod larvae. In the absence of hydromedusae or postlarval/juvenile fish, cohorts of larvae grew fast and survival rates were high. Cannibalism or predation by older cohorts on younger cohorts often brought the latter to extinction, even under favorable feeding conditions for the young larvae. Winter flounder larvae that were reared for two weeks in a 505 μm mesh, 11.5 m^3 enclosure in an estuary, had a survival rate of 76.8% (Laurence et al., 1979). These authors also believed that exclusion of predators resulted in high survival and implied that predation was a major regulator of winter flounder larval abundance.

Øiestad (1985) reached four conclusions based on his experience and review of mesocosm research:

(1) marine fish larvae have a very high potential survival rate, even at marginal feeding conditions;
(2) fast-growing and healthy larvae may be very susceptible to predation;
(3) from metamorphosis onward, larvae of some fish species are voracious predators of other fish larvae; and
(4) predation involves not only the prey that are ingested but also those that succumb to injuries inflicted by predators.

IV. Reproductive Strategies and Predation

If predation on eggs and larvae is intense and predictable, fishes should have adopted reproductive strategies to relieve predation pressure (Orians and Janzen, 1974; Johannes, 1978) or to compensate for it. Some possible strategies are listed by Johannes (1978), to which we add several others (Table 4). Attributing the evolution of these characteristics to the selective force of predation is speculative and some of these probably have evolved primarily for other reasons (Shapiro *et al.*, 1988). Johannes (1978) postulated that intense predation pressure on tropical marine fishes favors a reproductive strategy that is fundamentally different from that of fishes inhabiting higher latitudes. Tropical fishes have protracted spawning seasons in which eggs are spawned offshore or transported quickly offshore. Larvae may be returned later by favorable currents or eddies (Leis, 1982). Johannes (1978) believed that the offshore strategy avoids intense predation that would occur in coral reefs, seagrass meadows and mangrove estuaries. Because potential foods of larvae generally would be more abundant inshore, predation pressure is presumed to be the selective force for the observed spawning strategy. Individuals of tropical species may spawn repeatedly during the year. Such behavior could combat high and variable predation pressure on eggs and larvae as well as variable larval food supplies that might cause starvation.

Many reproductive strategies of tropical fishes are not useful at high latitudes and are not common there. For example, in the tropics spawning at night may be a "safe time", an adaptation to relieve predation pressure on pelagic eggs. In tropical waters, where hatching may take only 18–48 h, night spawning can significantly reduce the time that eggs are vulnerable to visual predators. However, at high latitudes where hatching typically takes from one to three weeks, spawning at night offers little advantage, except to provide a few hours for diffusion and dispersal of newly-spawned eggs. Holt *et al.* (1985) found that Gulf of Mexico sciaenid fishes typically spawn at

TABLE 4. REPRODUCTIVE STRATEGIES AND EGG AND LARVAL CHARACTERISTICS ADAPTED TO REDUCE PREDATION INTENSITY (*From Johannes, 1978)

Character/Strategy	Advantage	Examples, Prevalence
1. Spawning near or over deep water, offshore*	Removes prey from nearshore zone where predation is intense	Common in tropics
2. Spawning pelagic eggs at night*	Reduces exposure to visual predators	Common in tropics
3. Hatching demersal eggs at night*	Same as 2	Common in tropics
4. Release of pelagic eggs timed with outgoing tides*	Same as 1	Common in tropics
5. Hatching demersal eggs timed with outgoing tides*	Same as 1	Common in tropics
6. Spawning near shelter*	Protection for larvae	Several tropical species
7. Demersal spawning with egg guarding*	Protection for eggs	Common in tropics, littoral species, greenlings
8. Parental guarding of larvae*	Same as 6	Few species in tropics
9. Live bearing*	No egg predation, well-developed larvae	Few species in tropics Sebastes spp., elasmobranchs
10. Deep vertical distribution of eggs and larvae	Reduces predation by planktivores	Common, blue whiting, Pacific hake, halibut
11. Hard egg shell	Same as 7	Most species
12. Yolk-sac larvae motionless	Reduces predation by visual/ mechanoreceptive predators	Many species
13. Transparent eggs/yolk-sac larvae	Reduces predation by visual predators	Most species
14. Large eggs and larvae	Reduces predation by small invertebrates	Common, halibut
15. Fast growth/development rate	Reduces duration of vulnerable stages	Common, Scomber spp.
16. Patchy distribution	Reduces encounter probability, may swamp predators	Common
17. Short spawning season	May swamp predators, eliminate specialists	Common in subarctic
18. Spawning demersal eggs	Same as 10	Common in subarctic, Capelin, herring
19. Toxins in eggs or larvae	Same as 6 and 7	Uncommon, Fugu spp., cabezon
20. Batch spawning	Same as 16	Common in temperate regions

night and that overnight dispersal of eggs could be an important mechanism to reduce predation by fishes.

The distribution of eggs and larvae in water masses that are relatively free of predators (i.e. "safe sites") has been discussed by Alvariño (1980) for northern anchovy and later by Frank and Leggett (1985) and Leggett (1986) for capelin. Selection of "safe sites" by adult spawners decreases the probability of predation and also may increase feeding opportunities for larvae. Some species, for example blue whiting *Micromesistius poutassou*, spawn at very great depths (Coombs and Hiby, 1979) where predation pressure by planktivores on slowly ascending, developing eggs may be relatively low.

Mass spawning or synchronized hatching in response to environmental cues could satiate predators feeding on eggs and larvae and ultimately lead to higher survival rates. Frank and Leggett (1982) noted this phenomenon in capelin and suggested that hatching synchrony was

(1) cued to environmental conditions when predator densities were low and

(2) probably resulted in satiation of predators that were present.

Such synchrony also may have evolved to coincide with larval food production cycles.

Patchiness of eggs and larvae may reduce predation, but the presumed benefits remain speculative (Matsushita *et al.*, 1982; Heath and MachLachlan, 1985). Indeed, McGurk (1986) has proposed that the high mortality rates of fish eggs and larvae are a consequence of their patchy distributions. In theory, prey can reduce the probability of encounter with predators by aggregating, and predators that encounter patches may have difficulty targeting on individual prey (Hobson, 1978). Depending on the size of the patch and the density of prey within it, predators encountering a patch may become satiated quickly (Frank and Leggett, 1982) allowing most eggs or larvae to escape. But, if the predators are abundant or become satiated only after large consumption, the probable case for pelagic fishes, they could devastate the patch. Pelagic fishes in most marine habitats usually spawn batches of pelagic eggs. Some batches will be located and consumed but the net effect of temporal and spatial patchiness presumably will increase overall probability of survival.

McGurk (1986) reported that mortality rates of eggs and larvae increased as patchiness increased and believed that the higher mortality was from predation. McGurk's (1986) model was derived from data in the literature. He reported that Peterson and Wroblewski's (1984) model of marine organism mortality as a function of weight underestimated mortality rates of fish eggs and larvae (Fig. 15).

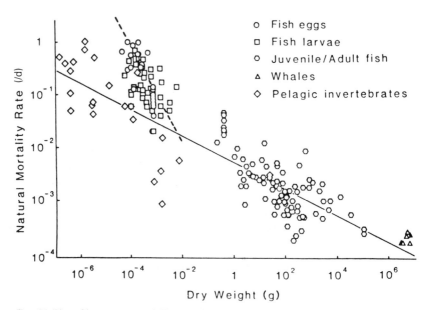

FIG. 15. Plot of instantaneous daily mortality rates on dry weight for marine organisms. See McGurk (1986) for data sources. Solid line is the mortality rates predicted by Peterson and Wroblewski's (1984) model. Broken line is the linear regression of log M on log W for fish eggs and larvae (from McGurk, 1986).

V. Fluctuations in Abundance, Control and Regulation

Predation is one of several processes that influence survival prior to recruitment (Fig. 16). However, it is difficult to assign egg or larval mortalities categorically to predation, starvation or physical processes. For example, starving larvae may be quickly preyed upon and thus eliminated from the environment. In this case the proximate cause of death is predation but the ultimate cause is nutrition-related. There is solid evidence that predation plays a decisive role in mortality of all early life stages but better experimental and analytical methods are needed to estimate not only mortalities induced by each major factor but also interaction effects.

Two of the probable major causes of early life mortality, starvation and predation, are closely linked by the growth and development process. Fast growth rates that occur under good feeding conditions can significantly reduce the time that young stages are exposed to high predation rates. This phenomenon, referred to as the "single process" by Cushing (1975a), has been discussed frequently (e.g. Ware, 1975; Shepherd and Cushing, 1980;

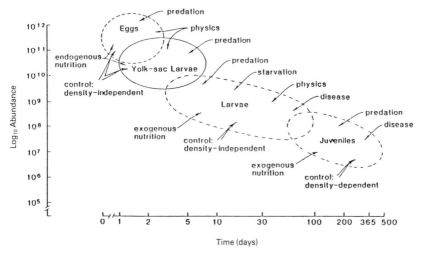

Fig. 16. A conceptualization of the recruitment process in fishes including the sources of nutrition, probable sources of death and hypothesized mechanism of control for four early life stages. Note that predation acts on all stages (from Houde, 1987).

Werner and Gilliam, 1984). Houde (1987) summarized growth-rate and larval stage duration variability for five species, and demonstrated how such variability could affect recruitment potential (Table 5). At constant predation rates, growth rates, which are influenced by available food and temperature, potentially could induce 100-fold or greater variability in survival to the juvenile stage. If predation rates are higher than the conservative mortality coefficients in Table 5, then even greater potential exists to induce recruitment fluctuations via growth rate variability. All mortality factors operate simultaneously but predation always will be a direct agent of mortality. Its ultimate impact can be modulated by surprisingly subtle differences in food levels and temperature. It is not necessary to postulate catastrophic losses owing directly to starvation or anomalous larval drift to explain year-class variability (Ware, 1975). It also is not necessary to infer that processes are either density-dependent or density-independent to explain fluctuations, although compensation must occur in some stages if populations are to persist with a degree of stability. This seems especially true considering the great potential for fluctuations caused by subtle events in early life.

The question of whether starvation or predation is more important as a cause of early life mortality remains unresolved and there may be no unequivocal answer because the situation may vary with species, area and year. In a recent review, Houde (1988) proposed that larval fishes from low latitudes are more likely to starve than those from high latitudes, based on

TABLE 5. PREDICTED NUMBERS OF SURVIVORS AT THE LOWEST AND HIGHEST PROBABLE LARVAL STAGE GROWTH RATES FOR FIVE SPECIES. THE RATIO OF HIGHEST NUMBER OF SURVIVORS (N_U) TO LOWEST NUMBER (N_L) IS AN ESTIMATE OF THE MAXIMUM POTENTIAL EFFECT ON RECRUITMENT THAT COULD BE CAUSED BY VARIABILITY IN LARVAL GROWTH RATE, GIVEN A FIXED MORTALITY RATE (From Houde, 1987)

Species	Instantaneous growth rate[a] (day⁻¹)		Larval stage duration[b] (day)		Instantaneous mortality coefficient (day⁻¹)	Number of survivors[c]		Ratio N_U:N_L
	G_L	G_U	T_L	T_U		N_L	N_U	
Bay Anchovy	0.15	0.35	50	21	0.18	123	22 823	185.6
Herring	0.03	0.10	173	52	0.04	988	124 930	126.4
Striped Bass	0.07	0.20	64	22	0.16	36	29 599	822.2
French Grunt	0.24	0.41	20	12	0.16	40 762	146 607	3.6
Cod	0.05	0.15	112	37	0.08	128	51 819	404.8

[a]G_L—Lowest probable rate; G_U—highest probable rate; [b]T_L—larval stage duration predicted for G_L; T_U—larval stage duration predicted for G_U; [c]N_L—number of survivors predicted for G_L; N_U—number of survivors predicted for G_U.

energetic considerations, especially the need to grow fast in tropical seas. It seems clear, however, that there are "prey-sensitive" and "prey-insensitive" species which, as larvae, are more or less likely to starve under food-limiting conditions. For example, anchovy larvae are "prey-sensitive" and they have a high probability of dying from food limitation at food concentrations that occur in the sea, but striped bass (*Morone saxatilis*) larval mortality rate is relatively insensitive to prey concentrations (Chesney, 1986; Houde, 1987). For bay anchovy, there is a critical range of food levels below which starvation will be the most likely cause of mortality. Within and above that range the probability of mortality shifts toward causes other than starvation, predation presumably being the major cause. For striped bass larvae, there is no critical or transition range of food among the levels likely to be encountered in Chesapeake Bay estuaries (Chesney, 1986). High rates of mortality reported for larval striped bass probably resulted from other factors, predation being a likely cause (Houde, 1987).

Lasker (1975, 1978) argued convincingly that northern anchovy larvae starve under unstable oceanographic conditions when food particles are dispersed. The "ocean stability" hypothesis is further supported by recent analyses which show that daily mortality declines under calm ocean conditions when food is aggregated to allow optimum anchovy larval feeding (Peterman and Bradford, 1987). It also is supported by recent work on cod and haddock (Buckley and Lough, 1987). But, Hunter (1982) found no evidence of a critical period at yolk-sac absorption for northern anchovy and suggested that the high, sustained mortality rate from egg to larval stages was caused primarily by predation. Bay anchovy larvae, despite being "prey-sensitive", sustained higher mortality from predation or other causes, than from starvation in Biscayne Bay (Leak and Houde, 1987).

Several lines of reasoning lead us to believe that predation may be the most important component of egg and larval mortality. First, with some notable exceptions (e.g. Theilacker, 1986; Buckley and Lough, 1987), there is little evidence that high proportions of larvae in the sea are starving (O'Connell, 1980; Sissenwine, 1984). Second, a marked increase in larval mortality due to starvation during the "critical period" has not been observed (May, 1974), although Fortier and Leggett (1985) have indicated that capelin mortality may increase significantly at the end of the yolk-sac stage when starvation could be a factor. Third, the abundance of potential predators in the sea is much higher than the abundance of fish eggs and larvae (McGowan and Miller, 1980; Kawai and Isibasi, 1983; Brewer *et al.*, 1984). Fourth, highest rates of mortality have been estimated for egg and pre-feeding, yolk-sac larval stages, which presumably have sufficient yolk reserves to resist starvation (Lo, 1986; Leak and Houde, 1987). Fifth, research in mesocosms, in the absence of predators, has demonstrated that larvae grow and survive at levels of food that are low compared with laboratory estimates of required amounts (Houde, 1985; Øiestad, 1985). Inclusion of predators in mesocosms devastates larval populations. A sixth line of evidence, based on recent experimental rearing, indicates that densities of food similar to those in the sea supported good growth and survival of larval Atlantic herring (Kiørboe and Munk, 1986), striped bass (Chesney, 1986) and other species (Houde and Schekter, 1981), indicating that starvation is less likely to be a direct cause of mortality than believed previously.

Studies of single predator populations indicate that predation may cause mortality rates ranging from 0.1%/day to 50%/h, depending on predator and prey characteristics (see Section IIIB). Recent estimates, where predation mortality is determined from total and starvation mortality, indicate that predation rates are very high: 50–80%/day for eggs and yolk-sac larvae of jack mackerel (Hewitt et al., 1985) and an average 86.5%/day for bay anchovy eggs (Leak and Houde, 1987). If predation is a major cause of egg and larval mortality, relatively small changes in predation rates can induce large recruitment variability. Small, but significant, changes in predation mortality could originate from variations in

(1) growth or development rate of larvae and eggs,
(2) distributions which may be biologically or physically mediated, causing variations in predator-prey overlap,
(3) abundance of predators, and
(4) availability of alternative prey for predators.

In freshwater systems, predation is recognized as an important force in structuring communities (Zaret, 1980). Crowder (1980) discussed the relationship between increases in two introduced pelagic species, rainbow smelt *Osmerus mordax* and alewife *Alosa pseudoharengus*, to reductions and extinctions of native fish populations in Lake Michigan. He concluded that predation on eggs and larvae by adult smelt and alewives may have caused collapses in populations of emerald shiner *Notropis atherinoides*, lake herring *Coregonus artedii*, and kiyi *C. kiyi* or fluctuations in populations of bloater *C. hoyi*, yellow perch *Perca flavescens* and sculpins *Cottus* spp. But, other authors have concluded that smelt predation on bloater or lake herring larvae had little effect on stocks in Lake Michigan and Lake Superior (Selgeby et al., 1978; Stedman and Argyle, 1985). Loftus and Hulsman (1986) presented convincing data and a simulation model that implicate rainbow smelt predation on larvae of whitefish *Coregonus clupeaformis* and lake herring *C. artedii* as the probable cause of collapse of those stocks in Twelve Mile Lake, Ontario. Other recent reports on predation effects on fish populations in lakes are Brandt et al. (1987) who observed significant predation by alewife on yellow perch larvae, and Lyons and Magnuson (1987) who noted predation by walleyes on larvae of littoral zone fishes.

In the sea, predation on eggs and larvae has not been demonstrated to be correlated with recruitment levels despite strong evidence that it is a major source of mortality. Knowledge of predator consumption rates and abundances indicates that fluctuations in year-class strength certainly could be induced by either density-independent or density-dependent predation on early life stages. It also seems likely that a regulatory role, through dampening of recruitment variability, can be ascribed to predation, particularly through density-dependent control during late postlarval or juvenile stages.

Spawner-recruit models of Ricker (1954) and Beverton and Holt (1957) are based on the concept of compensatory mortality (increasing mortality with increasing density) of pre-recruits. In the Ricker domed spawner-recruit model, the causes of density-dependent mortality are believed to result from aggregation of predators or cannibalism (Cushing, 1975b; MacCall, 1980). The Beverton–Holt model and more recent models link density-dependent growth with compensatory mortality (Jones, 1973; Cushing, 1974; Ware, 1975; Shepherd and Cushing, 1980). In these models it is assumed that increasing abundances of larvae result in competition for food and reduced growth rates, leading to prolonged larval stage duration and prolonged vulnerability to high predation rates. Conversely, when larval densities are low, it is assumed that competition for limited food resources is relaxed, growth rates approach maximum, and duration of the predator-vulnerable larval stage decreases, leading to better potential recruitment. These arguments remain controversial. Murphy (1961), Laurence (1982) and Cushing (1983) concluded that early-stage larvae are probably too dilute to affect the abundance of their prey. However, Bollens (1988) calculated that in Dabob Bay in Puget Sound, hake and herring larvae could affect their copepod prey abundance, and Kiørboe et al. (1988) suggested that density-dependent growth and survival was a feature of cohorts of autumn-spawned larval herring. In our opinion, density-dependent growth in early stages seems unlikely, except in cases of unusually high larval abundance. Environmental factors that affect larval growth rate can have a major influence on recruitment variability but probably do not play a density-dependent regulatory role.

Year-class fluctuations in plaice are believed to be determined in the egg and youngest larval stages (Harding et al., 1978; Veer, 1986). High recruitment levels are associated with cold spawning seasons. The main agent of mortality on eggs and larvae is thought to be predation (Harding et al., 1978) which operates in the 15 to 30-day egg stage. We have examined the 11 years of egg production and egg-larval mortality rates tabulated by Harding et al.

and have plotted mortality rates on estimated numbers of spawned eggs (Fig. 17). There is no evidence of density-dependent mortality in the egg stage and little evidence at larval stage I. At least fivefold variation in daily mortality rates occurred, which might be predation-related, but at this life stage, the effects apparently are density-independent.

Some characteristics of predation and probable population responses are

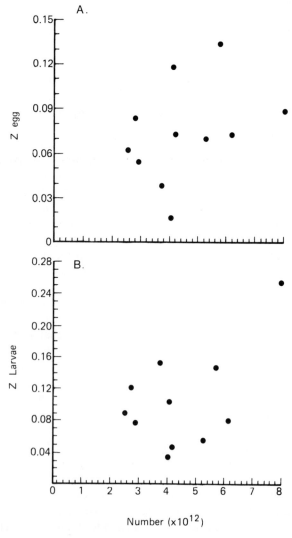

FIG. 17. Instantaneous mortality rate of (A) eggs and (B) early larvae of plaice plotted against the estimated number of spawned eggs (from data in Harding *et al.*, 1978).

summarized in Table 6. Direct causes of mortality other than predation are not considered here, although interacting factors associated with starvation or death from physical causes are listed to indicate potential for interactive effects. Coarse controls of a density-independent nature probably operate on the egg through larval stages, potentially producing large variability in year-class success. The possibility for regulation by density-dependent processes increases in the late post-larval through pre-recruit stages. This constitutes a fine-tuning of population responses to predation pressure.

In theory density-dependent regulation might occur at any stage, including egg (MacCall, 1980; Valdes et al., 1987), larval (Shepherd and Cushing, 1980; Crecco and Savoy, 1987), or juvenile (Lockwood, 1980; Sissenwine, 1984). MacCall (1980) believed that cannibalism by northern anchovy on eggs is an important regulatory mechanism that depends on adult-stock density and could produce a Ricker-type (dome-shaped) spawner-recruit curve. However, he concluded that density-independent factors tend to mask the relatively weak density-dependent influences. Further evidence that density-independent factors may conceal density-dependent components of mortality is provided by Welch (1986), who used statistical filtering to remove density-independent factors affecting recruitment in 16 stocks of eight species and improved the resulting stock-recruitment relationships in all cases. Welch concluded that previously obscured density-dependent, pre-recruit mortality was operating in all stocks.

The question of whether predation is a major cause of fluctuations and a regulator of year-class strength is further clouded by two other issues. These relate to life stage-specific effects and our lack of understanding of the functional and numerical responses of predators to prey density.

Population stabilization through compensatory predation mortality of late larval or juvenile fishes is possible if predator abundance is constant and they have a sigmoid (type III) functional response to prey density. Many vertebrates, including fishes, appear to have this type of functional response (Murdoch, 1969; Peterman and Gatto, 1978). Functional responses alone need not determine if predation stabilizes prey populations. But, numerical responses and prey switching behavior also may occur, generating increased mortality as prey density increases (Murdoch, 1969; Murdoch et al., 1975).

We believe, as does Pepin (1987), that in the sea predation losses of eggs and larvae are likely to be density-independent because most predators probably consume them only incidental to other, more common plankton organisms. Exceptions to this generalization will occur when predators aggregate on patches of prey, and perhaps when predators are selective and demonstrate density-dependent prey-switching behavior. In the late larval and juvenile stages many fish become the specific prey of piscivorous fishes. Under these circumstances mortality can be compensatory and year classes regulated, providing that predator capacity is not saturated.

TABLE 6. HYPOTHESIZED CHARACTERISTICS OF PREDATION AND POPULATION RESPONSES TO IT FOR FIVE LIFE STAGES OF MARINE FISHES

Life stage	Interacting factors	Mode of action	Regulatory potential	Probable recruitment implication
Egg	Water currents* Temperature Development rate Abundance of predators	Density-independent	Coarse	Year-class variations/ fluctuations
Yolk-sac larva	(Same as eggs)	Density-independent	Coarse	(Same as eggs)
Feeding larva	Water currents* Temperature Food supply Growth rate Abundance of predators	Probably Density-independent	Coarse	(Same as eggs)
Late larva/ Early juvenile	Food supply Growth rate Abundance of predators Predator functional response	Density-dependent (A) Compensatory (B) Depensatory	Coarse to fine	Year-class regulation, dampening of variability Year-class variations. Strong year-classes when initial numbers high; failed year-classes when initial numbers low
Pre-recruit	Growth rate Abundance of predators	Density-dependent	Fine	Year-class regulation; dampening of variability

*Here we include other physical factors, including fine to mesoscale turbulence and frontal dynamics.

Even though small changes in predation rate or duration of the predation process may cause major differences in recruitment level, it is often impossible to correlate larval abundances and resulting year-class size (Sissenwine, 1984; Peterman *et al.*, 1988) suggesting that regulation occurs after the larval stage. Sometimes these observations may be due to design and sampling problems, and rates rather than abundances may be more revealing. However, in some freshwater systems, such as Oneida Lake (New York), predation on juveniles rather than larvae has been demonstrated to be the major determinant of recruitment level (Forney, 1971; 1977). Peterman and Gatto (1978) found that the predatory capacity in salmon streams was seldom saturating and that type III functional responses (Holling, 1959) by fish predators on salmon fry generally prevailed. Where data were available, predation was a density-dependent, compensatory process. In theory, only at very high salmon fry densities would percentage mortality decline and the process become depensatory (Fig. 18). Forney (1971, 1977) reported that

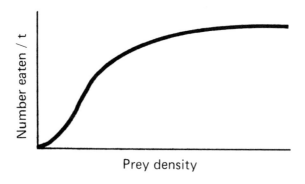

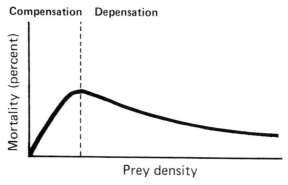

FIG. 18. Shapes of the type III functional response curve (top) and corresponding mortality curve (bottom) for increasing prey density (redrawn from Peterman and Gatto, 1978).

mortality of O^+ yellow perch from walleye (*Stizostedian vitreum*) predation in Oneida Lake was depensatory; the largest proportional mortalities were recorded for year-classes with smallest O^+ abundances. This suggests that the predatory capacity of walleye approaches saturation at the observed levels of prey abundance (Fig. 19). In contrast, cannibalism of O^+ yellow perch in Oneida Lake was compensatory (Tarby, 1974). Recently, Wood (1987) found that mortality of salmon fry from bird predation was depensatory because the birds are swamped by abundant prey.

For marine fishes, the stage about which we know least is the late-larval/early juvenile. Sissenwine (1984) argued that relative year-class strengths are determined by predation at this stage. We think that determination of year-class size for most marine fishes is probable at earlier stages but acknowledge that coarse to fine adjustment by predation in the late-larval/juvenile stage can have moderate to large effects on recruitment potential. This may be especially true for species such as cod or herring which have slow growth rates and long late-larval/juvenile stage durations, or perhaps for some tropical reef fishes that have exceptionally short larval stage durations (Houde, 1987). Potential variability in larval stage duration also is much greater in high latitude seas than in the tropics (Houde, 1988), leading to the possibility that density-dependent regulation is more likely to occur in high latitudes where predators have time to aggregate and where competitive relationships have time to develop during an extended larval stage.

Veer and Bergman (1987) found that predation by the brown shrimp *Crangon crangon* caused a significant density-dependent mortality of recently metamorphosed plaice juveniles. This predation appeared to account for the total mortality of plaice juveniles from February to May. The authors suggested that density-dependent control was due to combined effects of the functional and numerical responses of shrimp to plaice density. They attributed coarse control of year-class strength to variations in egg and larval survival, with subsequent shrimp predation acting as a fine control to reduce between-year variability.

High predation rates at the late-larval/juvenile stage do not necessarily regulate year-class strength through compensatory controls. Research on both marine and freshwater fishes has demonstrated that strong year-classes can emerge despite high predation rates on juveniles (Bailey and Ainley, 1983; Dwyer *et al.*, 1987; Forney, 1971, 1976). In such cases, depensatory predation mortality is thought to contribute to dominant year-class production in years when survival to the late-larval stage was high. Under conditions of high-larval survival but subsequent compensatory mortality, small year classes could result. If survival of eggs and larvae were low, predation on late-larval/early juvenile fish theoretically could contribute to

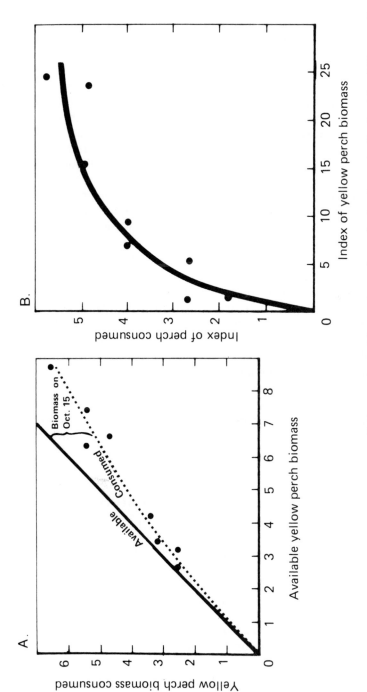

Fig. 19. (A) Relation between biomass of perch available to predators (biomass August 1 + production August 1 – October 15) and biomass of perch consumed in absence of other mortality. (B) Relative weight of perch consumed by the walleye population between July and September and index of juvenile perch biomass on August 15, 1959–1966 estimated from trawl catches (redrawn from Forney, 1971).

year-class failure, although the coarse controls that caused mortality in younger stages would be primarily responsible.

These scenarios are supported by Smith's (1985) simulation studies of northern anchovy where he examined the sensitivity of recruitment to changes in the parameters of mortality, growth and stage duration. Moderate changes in egg and larval mortality could easily amount for five-fold increases or decreases in recruitment. Although a moderate increase in juvenile mortality could account for a five-fold decrease in anchovy recruitment, no reasonable change in juvenile mortality rate could cause a five-fold increase.

The nature of stage-specific predation in the recruitment process may vary among regions, although the causes and consequences are speculative at this time. In any case, fishes from low latitudes apparently have evolved numerous mechanisms to reduce predation on egg and larval stages (Johannes, 1978; Holt et al., 1985; but see Shapiro et al., 1988) which may be a consequence of increased hunger and metabolic rates of predators at high temperatures (Pauly, 1980). In upwelling and other mid- to low latitude regions daily spawning by adults and fast growth of egg and larval stages can mitigate predation losses. The toll on eggs and larvae still could be high because of their consistent abundance in the plankton and the evidence that some predators or cannibals feed specifically on eggs and larvae (MacCall, 1980; Purcell, 1985). In high latitudes eggs and larvae occur in a relatively short season but grow slowly during that time, increasing individual vulnerability to predation. At the same time, growth of late-larvae/juveniles also is relatively slow in high latitudes, and large size relative to predators, which should afford protection, may not be achieved until many months after hatching.

VI. Directions for the Future

Productive research in the coming years will include:

Field programs to examine egg/larval mortality and predation on space and time scales appropriate to the processes involved, as suggested also by Leggett (1986). We believe that such studies can be most rewarding in high latitudes where the diversity of eggs/larvae and their predators is relatively low.

Better estimates of total mortality rates in early life stages from which starvation and transport loss components could be subtracted to obtain an estimate of predation loss. This may be the only realistic approach to estimate the extent of predation mortality in high diversity, low latitude ecosystems. It will require an interdisciplinary effort, new analytical procedures and the development of models to predict major age-specific agents of mortality.

Experimental approaches in mesocosms to study predation processes. Specific predators and their stage-specific effects can be investigated. This approach will allow models of predator–prey interactions to be developed. It also offers the best opportunity to concurrently examine how predation, starvation and anthropogenic influences interact. Cautious interpretations will be necessary because of confinement effects, lack of advection and reduced turbulence. Both short-term (hours to days) and long-term (e.g. entire larval stage) experiments are possible.

Defining the role of predation and cannibalism by fishes on the late-larval/ early juvenile stage. Fishes may be the single most important group of predators on early life stages. Field and experimental programs are essential to investigate the role of fishes as stage-specific predators and to determine their potential to regulate year-class size or cause major fluctuations in abundance.

All of the recommended research will require long-term efforts and must be well-integrated with research on stock-recruitment relationships and environmental variability. Predation research also must be increasingly interdisciplinary because of the diverse factors that shape the recruitment process in the early life of fishes.

Acknowledgements

We thank J. Blaxter, A. Kendall, J. Purcell, and W. Palsson for comments on the manuscript and R. Olson, A. Mullin, R. Peterman, R. Brodeur and G. Theilacker for comments on specific sections. We also thank J. Yen and J. Purcell for access to unpublished material and manuscripts.

References

Alheit, J. (1987). Egg cannibalism versus egg predation: their significance in ancho-vies. *South African Journal of Marine Science* **5**, 467–470.
Ali Kahn, J. and Hempel, G. (1974). Relation of fish larvae and zooplankton biomass in the Gulf of Aden. *Marine Biology* **28**, 311–316.
Alvariño, A. (1980). The relation between the distribution of zooplankton predators and anchovy larvae. *California Cooperative Oceanic Fisheries Investigation Reports* **20**, 150–160.
Alvariño, A. (1981). The relation between the distribution of zooplankton predators and anchovy larvae. *Rapports et Procès-Verbaux des Réunions. Conseil Internat-ional pour l'Exploration de la Mer* **178**, 197–199.
Alvariño, A. (1985). Predation in the plankton realm; mainly with reference to fish larvae. *Investigaciones Marinas Centro Interdisciplinario de Ciencias Mariana* **2**, 1–122.

Arai, M. N. and Hay, D. E. (1982). Predation by medusae on Pacific herring (*Clupea harengus pallasi*) larvae. *Canadian Journal of Fisheries and Aquatic Sciences* **39**, 1537–1540.
Bailey, K. M. (1982). The early life history of Pacific hake. *Fishery Bulletin U.S.* **80**, 589–598.
Bailey, K. M. (1984). Comparison of laboratory rates of predation on five species of marine fish larvae by three planktonic invertebrates: effects of larval size on vulnerability. *Marine Biology* **79**, 303–309.
Bailey, K. M. and Ainley, D. G. (1982). The dynamics of California sea lion predation on Pacific hake. *Fisheries Research* **1**, 163–176.
Bailey, K. M. and Batty, R. S. (1983). A laboratory study of predation by *Aurelia aurita* on larval herring (*Clupea harengus*): experimental observations compared with model predictions. *Marine Biology* **72**, 295–301.
Bailey, K. M. and Batty, R. S. (1984). Laboratory study of predation by *Aurelia aurita* on larvae of cod, flounder, plaice and herring: development and vulnerability to capture. *Marine Biology* **83**, 287–291.
Bailey, K. M. and Stehr, C. L. (1986). Laboratory studies on the early life history of the walleye pollock, *Theragra chalcogramma* (Pallas). *Journal of Experimental Marine Biology and Ecology* **99**, 233–246.
Bailey, K. M. and Yen, J. (1983). Predation by a carnivorous marine copepod, *Euchaeta elongata* Esterly, on eggs and larvae of the Pacific hake, *Merluccius productus*. *Journal of Plankton Research* **5**, 71–81.
Bajkov, A. D. (1935). How to estimate the daily food consumption of fish under natural conditions. *Transactions of the American Fisheries Society* **65**, 288–289.
Bakun, A., Beyer, J., Pauly, D., Pope, J. G. and Sharp, G. D. (1982). Ocean sciences in relation to living resources. *Canadian Journal of Fisheries and Aquatic Sciences* **39**, 1059–1070.
Bartell, S. M., Breck, J. E., Gardner, R. H. and Brenkert, A. L. (1986). Individual parameter perturbation and error analysis of fish bioenergetics. *Canadian Journal of Fisheries and Aquatic Sciences* **43**, 160–168.
Batty, R. S. (1981). Locomotion of plaice larvae. *Symposium of the Zoological Society of London* **48**, 53–69.
Batty, R. S. (1984). Development of the swimming movements and musculature of larval herring (*Clupea harengus*). *Journal of Experimental Biology* **110**, 217–229.
Batty, R. S. (1987). Effect of light intensity on activity and food-searching of larval herring, *Clupea harengus*: a laboratory study. *Marine Biology* **94**, 323–327.
Batty, R. S., Blaxter, J. H. S. and Libby, D. A. (1986). Herring (*Clupea harengus*) filter-feeding in the dark. *Marine Biology* **91**, 371–375.
Beverton, R. J. H. and Holt, S. J. (1957). On the dynamics of exploited fish populations. U.K. Ministry of Agriculture and Fisheries. Fisheries Investigations. (Series 2), 533 pp.
Blaxter, J. H. S. (1986). Development of sense organs and behaviour of teleost larvae with special reference to feeding and predator avoidance. *Transactions of the American Fisheries Society* **115**, 98–114.
Blaxter, J. H. S. (1988). Pattern and variety in development. *Fish Physiology* **11A**, 1–58.
Blaxter, J. H. S. and Batty, R. S. (1985). The development of startle responses in herring larvae. *Journal of the Marine Biological Association of the United Kingdom* **65**, 737–750.
Blaxter, J. H. S., Gray, J. A. B. and Best, C. G. (1983). Structure and development of the free neuromast and lateral line system of the herring. *Journal of the Marine*

Biological Association of the United Kingdom **63**, 247–260.

Boisclair, D. and Leggett, W. C. (1988). An *in situ* experimental evaluation of the Elliott and Persson and the Eggers models for estimating fish daily ration. *Canadian Journal of Fisheries and Aquatic Sciences* **45**, 138–145.

Bollens, S. M. (in press). A model of the predatory impact of larval marine fish on the population dynamics of their zooplankton prey. *Journal of Plankton Research.* **10**, 887–906.

Bowman, A. (1922). Spawny haddocks; the occurrence of spawny haddock and the locus and extent of herring spawning grounds. Fisheries Investigations of Scotland 1922 (No. 4), 15 pp.

Brandt, S. B., Mason, D. M., MacNeill, D. B., Coates, T. and Gannon, J. E. (1987). Predation by alewives on larvae of yellow perch in Lake Ontario. *Transactions of the American Fisheries Society* **116**, 641–645.

Brewer, G. D., Kleppel, G. S. and Dempsey, M. (1984). Apparent predation on ichthyoplankton by zooplankton and fishes in nearshore waters of southern California. *Marine Biology* **80**, 17–28.

Brooks, J. L. and Dodson, S. I. (1965). Predation, body size and composition of plankton. *Science* **150**, 28–35.

Brownell, C. L. (1985). Laboratory analysis of cannibalism by larvae of the cape anchovy *Engraulis capensis. Transactions of the American Fisheries Society* **114**, 512–518.

Brownell, C. L. (1987). Cannibalistic interactions among young anchovy: a first attempt to apply laboratory behavioural observations to the field. *South African Journal of Marine Science* **5**, 503–512.

Buckley, L. J. and Lough, R. G. (1987). Recent growth, biochemical composition and prey field of larval haddock (*Melanogrammus aeglefinus*) and Atlantic cod (*Gadus morhua*) on Georges Bank. *Canadian Journal of Fisheries and Aquatic Sciences* **44**, 14–25.

Butler, J. L. and Pickett, D. (1988). Age-specific vulnerability of Pacific sardine, *Sardinops sagax*, larvae to predation by northern anchovy, *Engraulis mordax. Fishery Bulletin U.S.* **86**, 163–167.

Chesney, E. J. (1986). Multiple environmental factors as determinants of survival and growth in larval striped bass, *Morone saxatilis.* International Council for the Exploration of the Sea, C.M. 1986/M: 26, 9 pp. (mimeo).

Christensen, V. (1983). Predation by sand eel on herring larvae. International Council for the Exploration of the Sea, C.M. 1983/L: 27, 10 pp. (mimeo).

Colin, P. L. (1976). Filter feeding and predation on the eggs of *Thallasoma sp.* by the scombroid fish *Rastrelliger kanagurta. Copeia 1976*, 596–597.

Confer, J. L. and Blades, P. I. (1975). Omnivorous zooplankton and planktivorous fish. *Limnology and Oceanography* **20**, 571–579.

Coombs, S. H. and Hiby, A. R. (1979). The development of the eggs and early larvae of blue whiting, *Micromesistius poutassou* and the effect of temperature on development. *Journal of Fish Biology* **14**, 111–123.

Cooper, S. D., Smith, D. W. and Bence, J. R. (1985). Prey selection by freshwater predators with different foraging strategies. *Canadian Journal of Fisheries and Aquatic Sciences* **42**, 1720–1732.

Corten, A. (1983). Predation on herring larvae by the copepod *Candacia armata.* International Council for the Exploration of the Sea C.M. 1983/H: 20, 3 pp. (mimeo).

Crecco, V. and Savoy, T. (1987). Effects of climatic and density-dependent factors on intra-annual mortality of larval American shad. *American Fisheries Society Symposium* **2**, 69–81.

Crowder, L. B. (1980). Alewife, rainbow smelt and native fishes in Lake Michigan: competition or predation? *Environmental Biology of Fishes* **5**, 225–233.

Curio, E. (1976). "The Ethology of Predation". Springer-Verlag, New York.

Cushing, D. H. (1972). The production cycle and the numbers of marine fish. *Symposium of the Zoological Society of London* **29**, 213–232.

Cushing, D. H. (1974). The possible density-dependence of larval mortality and adult mortality in fishes. *In* "The Early Life History of Fish" (J. H. S. Blaxter, ed.), pp. 103–111, Springer-Verlag, Berlin.

Cushing, D. H. (1975a). "Marine Ecology and Fisheries". Cambridge University Press, Cambridge, 278 pp.

Cushing, D. H. (1975b). The natural mortality of plaice. *Journal du Conseil, Conseil International pour l'Exploration de la Mer* **36**, 150–157.

Cushing, D. H. (1980). The decline of the herring stocks and the gadoid outburst. *Journal du Conseil, Conseil International pour l'Exploration de la Mer* **39**, 70–81.

Cushing, D. H. (1983). Are fish larvae too dilute to affect the density of their food organisms? *Journal of Plankton Research* **5**, 847–854.

Cushing, D. H. (1984). The gadoid outburst in the North Sea. *Journal du Conseil. Conseil International pour l'Exploration de la Mer* **41**, 159–166.

Daan, N. (1976). Some preliminary investigations into predation on fish eggs and larvae in the southern North Sea. International Council for the Exploration of the Sea, C.M. 1976/L: 15, 11 pp. (mimeo).

Daan, N., Rijnsdorp, A. D. and Overbeeke, G. R., van (1985). Predation by North Sea herring *Clupea harengus* on eggs of plaice *Pleuronectes platessa* and cod *Gadus morhua*. *Transactions of the American Fisheries Society* **114**, 499–506.

de Lafontaine, Y. and Leggett, W. C. (1987). Effect of container size on estimates of mortality and predation rates in experiments with macrozooplankton and larval fish. *Canadian Journal of Fisheries and Aquatic Sciences* **44**, 1534–1543.

de Lafontaine, Y. and Leggett, W. C. (1988). Predation by jellyfish on larval fish: an experimental evaluation employing *in situ* enclosures. *Canadian Journal of Fisheries and Aquatic Science* **445**, 1173–1190.

Dill, L. M. (1974a). The escape response of the zebra danio (*Brachydanio rerio*). I. The stimulus for escape. *Animal Behavior* **22**, 711–722.

Dill, L. M. (1974b). The escape response of the zebra danio (*Brachydanio rerio*). II. The effect of experience. *Animal Behavior* **22**, 723–730.

Dowd, C. E. (1986). Predator-prey interactions between juvenile bay anchovy (*Anchoa mitchilli*) and larvae of sea bream (*Archosargus rhomboidalis*). Ph.D. Dissertation, University of Miami, Coral Gables, Fla., 134 pp.

Dragesund, O. and Nakken, O. (1973). Relationship of parent stock size and year class strength in Norwegian spring herring. *Rapports et Procès-Verbaux des Réunions. Conseil International pour l'Exploration de la Mer* **164**, 15–29.

Durbin, A. G. and Durbin, E. G. (1975). Grazing rates of the Atlantic menhaden *Brevoortia tyrannus* as a function of particle size and concentration. *Marine Biology* **33**, 265–277.

Dwyer, D. A., Bailey, K. M. and Livingston, P. A. (1987). Feeding habits and daily ration of walleye pollock (*Theragra chalcogramma*) in the eastern Bering Sea, with special reference to cannibalism. *Canadian Journal of Fisheries and Aquatic Sciences* **44**, 1972–1984.

Eaton, R. C. and DiDomenico, R. (1986). Role of the telost escape response during development. *Transactions of the American Fisheries Society* **115**, 128–142.

Eaton, R. C., Farley, R. D., Kimmel, C. B. and Schabtach, E. (1977). Functional development in the Mauthner cell system of embryos and larvae of the zebra fish. *Journal of Neurobiology* **8**, 151–172.

Eaton, R. C. and Hackett, J. T. (1984). The role of the Mauthner cell in fast-starts involving escape in teleost fish. *In* "Neural Mechanisms of Startle Behavior" (R. C. Eaton, ed.), pp. 213–262. Plenum Press, N.Y.

Eggers, D. M. (1977). The nature of prey selection by planktivorous fish. *Ecology* **58**, 46–59.

Elliott, J. M. and Persson, L. (1978). The estimation of daily rates of food consumption for fish. *Journal of Animal Ecology* **47**, 977–991.

Fabian, M. W. (1960). Mortality of fresh water and tropical fish fry by cyclopoid copepods. *Ohio Journal of Science* **60**, 268–270.

Fancett, M. S. (1988). Diet and selectivity of scyphomedusae from Port Phillip, Australia. *Marine Biology* **98**, 503–509.

Fancett, M. S. and Jenkins, G. P. (1988). Predatory impact of scyphomedusae on ichthyoplankton and other zooplankton in Port Phillip Bay. *Journal of Experimental Marine Biology and Ecology* **116**, 63–77.

Feigenbaum, D. and Reeve, M. R. (1977). Prey detection in the Chaetognatha: response to a vibrating probe and experimental determination of attack distance in large aquaria. *Limnology and Oceanography* **22**, 1052–1058.

Feller, R. J., Taghon, G. L., Gallagher, E. D. and Jumars, P. A. (1979). Immunological methods for food web analysis in a soft-bottom benthic community. *Marine Biology* **54**, 61–74.

Folkvord, A. and Hunter, J. R. (1986). Size-specific vulnerability of northern anchovy, *Engraulis mordax*, larvae to predation by fishes. *Fishery Bulletin U.S.* **84**, 859–869.

Forney, J. L. (1971). Development of dominant year classes in a yellow perch population. *Transactions of the American Fisheries Society* **100**, 739–749.

Forney, J. L. (1976). Year-class formation in the walleye (*Stizostedion vitreum vitreum*) population of Oneida Lake, New York. *Journal of the Fisheries Research Board of Canada* **33**, 783–792.

Forney, J. L. (1977). Reconstruction of yellow perch (*Perca flavescens*) cohorts from examination of walleye (*Stizostedium vitreum vitreum*) stomachs. *Journal of the Fisheries Research Board of Canada* **34**, 925–932.

Fortier, L. and Leggett, W. C. (1985). A drift study of larval fish survival. *Marine Ecology Progress Series* **25**, 245–257.

Frank, K. T. (1986). Ecological significance of the ctenophore *Pleurobrachia pileus* off southwestern Nova Scotia. *Canadian Journal of Fisheries and Aquatic Sciences* **43**, 211–222.

Frank, K. T. and Leggett, W. C. (1982). Coastal water mass replacement: its effect on zooplankton dynamics and the predator-prey complex associated with larval capelin (*Mallotus villosus*). *Canadian Journal of Fisheries and Aquatic Sciences* **39**, 991–1003.

Frank, K. T. and Leggett, W. C. (1984). Selective exploitation of capelin (*Mallotus villosus*) eggs by winter flounder (*Pseudopleuronectes americanus*): capelin egg mortality rates, and contribution of egg energy to the annual growth of flounder. *Canadian Journal of Fisheries and Aquatic Sciences* **41**, 1294–1302.

Frank, K. T. and Leggett, W. C. (1985). Reciprocal oscillations in densities of larval fish and potential predators: a reflection of present or past predation? *Canadian Journal of Fisheries and Aquatic Sciences* **42**, 1841–1849.

Fraser, J. H. (1969). Experimental feeding of some medusae and chaetognatha. *Journal of the Fisheries Research Board of Canada* **26**, 1743–1762.

Fraser, J. H. (1970). The ecology of the ctenophore *Pleurobrachia pileus* in Scottish waters. *Journal du Conseil. Conseil International pour l'Exploration de la Mer* **33**, 149–168.

Fuiman, L. A. (1986). Burst-swimming performance of larval zebra danios and the effects of diel temperature fluctuations. *Transactions of the American Fisheries Society* **115**, 143–148.

Fuiman, L. A. and Gamble, J. C. (1988). Predation by Atlantic herring, sprat, and sandeels on herring larvae in large enclosures. *Marine Ecology Progress Series* **44**, 1–6.

Gamble, J. C. (1985). More space for the sparsely distributed: an evaluation of the use of large enclosed ecosystems in fish larval research. International Council for the Exploration of the Sea, C.M. 1985/Mini-Symposium No. 5 (mimeo).

Gamble, J. C., MacLachlan, P. and Seaton, D. D. (1985). Spaced-out larvae survive: a large-scale look at predation by *Aurelia aurita* on yolk-sac herring larvae. International Council for the Exploration of the Sea, C.M. 1985/Mini-symposium No. 14, 2 pp. (mimeo).

Garrod, C. and Harding, D. (1981). Predation by fish on the pelagic eggs and larvae of fishes spawning in the west central North Sea in 1976. International Council for the Exploration of the Sea, C.M. 1981/L:11 (mimeo).

Gerritsen, J. (1980). Adaptive responses to encounter problems. *In* "Evolution and Ecology of Zooplankton Communities" (W. C. Kerfoot, ed.), pp. 52–62, University Press of New England, Hanover, N.H.

Gerritsen, J. and Strickler, J. R. (1977). Encounter probabilities and community structure in zooplankton: a mathematical model. *Journal of the Fisheries Research Board of Canada* **34**, 73–82.

Gibson, R. N. and Ezzi, I. A. (1985). Effect of particle concentration on filter- and particulate feeding in the herring *Clupea harengus*. *Marine Biology* **88**, 109–116.

Gophen, M. and Harris, R. P. (1981). Visual predation by a marine cyclopoid copepod, *Corycaeus anglicus*. *Journal of the Marine Biological Association of the United Kingdom* **61**, 391–399.

Grave, H. (1981). Food and feeding of mackerel larvae and early juveniles in the North Sea. *Rapports et Procès-Verbaux des Réunions. Counseil International pour l'Exploration de la Mer* **178**, 454–459.

Greene, C. H. (1985a). Selective predation in pelagic communities. Ph.D. Dissertation, University of Washington, Seattle, 155 pp.

Greene, C. H. (1985b). Planktivore functional groups and patterns of prey selection in pelagic communities. *Journal of Plankton Research* **7**, 35–40.

Greene, C. H. (1986). Patterns of prey selection: Implications of predator foraging tactics. *American Naturalist* **128**, 824–839.

Gulland, J. A. and Garcia, S. (1984). Observed patterns in multispecies fisheries. *In* "Exploitation of Marine Communities" (R. M. May, ed.), pp. 155–190, Dahlem Konferenzen, Springer-Verlag, Berlin.

Hamner, P. and Hamner, W. N. (1977). Chemosensory tracking of scent trails by the planktonic shrimp *Acetes sibogac australis*. *Science* **195**, 886–888.

Harding, D., Nichols, J. H. and Tungate, D. S. (1978). The spawning of plaice (*Pleuronectes platessa* L.) in the southern North Sea and English Channel. *Rapports et Procès-Verbaux des Réunions. Conseil International pour l'Exploration de la Mer* **172**, 102–113.

Hartig, J. H. and Jude, D. J. (1984). Opportunistic cyclopoid predation on fish larvae. *Canadian Journal of Fisheries and Aquatic Sciences* **41**, 526–532.

Hartig, J. H. and Jude, D. J. (1988). Ecological and evolutionary significance of cyclopoid predation on fish larvae. *Journal of Plankton Research* **10**, 573–577.

Hartig, J. H., Jude, D. J. and Evans, M. S. (1982). Cyclopoid predation on Lake Michigan fish larvae. *Canadian Journal of Fisheries and Aquatic Sciences* **39**, 1563–1568.

Hattori, S. (1962). Predatory activity of *Noctiluca* on anchovy eggs. *Bulletin Tokai Regional Fisheries Research Laboratory* **9**, 211–220.

Heath, M. R. and MachLachlan, P. (1985). Growth and survival rates of yolk-sac herring larvae from a spawning ground to the west of the Outer Hebrides. International Council for the Exploration of the Sea, C.M. 1985/H:28, 14 pp. (mimeo).

Heeger, T. and Moller, H. (1987). Ultrastructural observations on prey capture and digestion in the scyphomedusa *Aurelia aurita*. *Marine Biology* **96**, 391–400.

Hewitt, R. P. (1981). The value of pattern in the distribution of young fish. *Rapports et Procès-Verbaux des Réunions. Conseil International pour l'Exploration de la Mer* **178**, 229–236.

Hewitt, R. P., Theilacker, G. H. and Lo, N. C. H. (1985). Causes of mortality in young jack mackerel. *Marine Ecology Progress Series* **26**, 1–10.

Hickey, G. M. (1979). Survival of fish larvae after injury. *Journal of Experimental Marine Biology and Ecology* **37**, 1–17.

Hickey, G. M. (1982). Wound healing in fish larvae. *Journal of Experimental Marine Biology and Ecology* **57**, 149–168.

Hildén, M. (1988). Significance of the functional response of predators to changes in prey abundance in multispecies virtual population analysis. *Canadian Journal of Fisheries and Aquatic Sciences* **45**, 89–96.

Hjort, J. (1914). Fluctuations in the great fisheries of Northern Europe. *Rapports et Procès-Verbaux des Réunions. Conseil International pour l'Exploration de la Mer* **20**, 1–13.

Hobson, E. S. (1978). Aggregating as a defense against predators in aquatic and terrestrial environments. *In* "Contrasts in Behavior" (E. S. Reese and F. J. Lighter, eds), pp. 219–234, John Wiley & Sons, Chichester, UK.

Holling, C. S. (1959). The components of predation as revealed by a study of small mammal predation of the European sawfly. *Canadian Journal of Entomology* **91**, 293–332.

Holt, G. J., Holt, S. A. and Arnold, C. R. (1985). Diel periodicity of spawning in sciaenids. *Marine Ecology Progress Series* **27**, 1–7.

Houde, E. D. (1985). Mesocosms and recruitment mechanisms. International Council for the Exploration of the Sea, C.M. 1981/Mini Symposium No. 4: 13 pp. (mimeo).

Houde, E. D. (1987). Fish early life dynamics and recruitment variability. *American Fisheries Society Symposium* **2**, 17–29.

Houde, E. D. (1988). Comparative growth and energetics of marine fish larvae. International Council for the Exploration of the Sea, Early Life History Symposium, ELHS No. 2, 20 pp. (mimeo).

Houde, E. D. and Schekter, R. C. (1981). Growth rates, rations and cohort consumption of marine fish larvae in relation to prey concentrations. *Rapports et Procès-Verbaux des Réunions. Conseil International pour l'Exploration de la Mer* **178**, 441–453.

Houde, E. D., Almatar, S., Leak, J. L. and Dowd, C. E. (1986). Ichthyoplankton abundance and diversity in the western Arabian Gulf. *Kuwait Bulletin of Marine Science* **8**, 107–393.

Hourston, A. S., Rosenthal, H. and Kerr, S. (1981). Capacity of juvenile herring to feed on larvae of their own species. *Canadian Technical Report Series. Fisheries and Aquatic Science* **1044**, 9 pp.

Hunt, G. L. and Butler, J. L. (1980). Reproductive ecology of western Gulls and Xantus murrelets with respect to food resources in the southern California Bight. *California Cooperative Oceanic Fisheries Investigation Reports* **21**, 62–78.

Hunter, J. R. (1972). Swimming and feeding behavior of larval anchovy. *Fishery Bulletin U.S.* **70**, 821–838.

Hunter, J. R. (1976). Report of a colloquium on larval fish mortality studies and their relation to fishery research, January 1975. *National Oceanic and Atmospheric Administration Technical Report, National Marine Fisheries Service Circular-395*, 5 pp.

Hunter, J. R. (1981). Feeding ecology and predation of marine fish larvae. *In* "Marine Fish Larvae: Morphology, Ecology and Relation to Fisheries" (R. Lasker, ed.) pp. 33–77, University of Washington Press, Seattle.

Hunter, J. R. (1982). Predation and recruitment. Fish Ecology III, *University of Miami Technical Report* **82008**, 172–209.

Hunter, J. R. (1984). Inferences regarding predation on the early life stages of cod and other fishes. *In* "The Propagation of Cod" (E. Dahl, D. Danielssen, E. Moksness and P. Solemdal, eds). *Flødevigen Rapportseries* **1**, 533–562.

Hunter, J. R. and Dorr, H. (1982). Thresholds for filter feeding in northern anchovy, *Engraulis mordax*. *California Cooperative Oceanic Fisheries Investigation Reports* **23**, 198–204.

Hunter, J. R. and Kimbrell, C. A. (1980). Egg cannibalism in the northern anchovy, *Engraulis mordax*. *Fishery Bulletin U.S.* **78**, 811–816.

Ivlev, V. S. (1961). "Experimental Ecology of the Feeding of Fishes." Yale University Press, New Haven, Connecticut.

Janssen, J. (1982). Comparison of searching behavior for zooplankton in an obligate planktivore, blueback herring (*Alosa aestivalis*) and a facultative planktivore, bluegill (*Lepomis macrochirus*). *Canadian Journal of Fisheries and Aquatic Sciences* **39**, 1649–1654.

Jobling, M. (1981). Mathematical models of gastric emptying and the estimation of daily rates of food consumption for fish. *Journal of Fish Biology* **19**, 245–257.

Jobling, M. (1986). Mythical models of gastric emptying and implications for food consumption studies. *Environmental Biology of Fishes* **16**, 35–50.

Johannes, R. E. (1978). Reproductive strategies of coastal marine fishes in the tropics. *Environmental Biology of Fishes* **3**, 65–84.

Johannessen, A. (1980). Predation on herring (*Clupea harengus*) eggs and young larvae. International Council for the Exploration of the Sea, C.M./1980, H: 33, 12 pp. (mimeo).

Jones, R. (1973). Density dependent regulation of the numbers of cod and haddock. *Rapports et Procès-Verbaux des Réunions. Conseil International pour l'Exploration de la Mer* **164**, 156–173.

Kawai, T. and Isibasi, K. (1983). Change in abundance and species composition of neritic pelagic fish stocks in connection with larval mortality caused by cannibalism and predatory loss by carnivorous plankton. *FAO Fisheries Report* **291**, 1081–1111.

Kerfoot, W. C. (1978). Combat between predatory copepods and their prey: *Cyclops*, *Epischura*, and *Bosmina*. *Limnology and Oceanography* **23**, 1089–1102.

Kimmel, C. B., Eaton, R. C. and Powell, S. L. (1980). Decreased fast-start performance of zebrafish lacking Mauthner neurons. *Journal of Comparative Physiology* **140**, 343–350.

Kiørboe, T. and Munk, P. (1986). Feeding and growth of larval herring, *Clupea harengus*, in relation to density of copepod nauplii. *Environmental Biology of Fishes* **17**, 133–139.

Kiørboe, T., Munk, P., Richardson, K., Christensen, V. and Paulsen, H. (1988). Plankton dynamics and larval herring growth, drift and survival in a frontal area. *Marine Ecology Progress Series* **44**, 205–219.

Kislalioglu, M. and Gibson, R. N. (1976). Some factors governing prey selection by the 15-spined stickleback, *Spinachia spinachia* L. *Journal of Experimental Marine Biology and Ecology* **25**, 159–169.

Koslow, J. A. (1984). Recruitment patterns in northwest Atlantic fish stocks. *Canadian Journal of Fisheries and Aquatic Sciences* **41**, 1722–1729.

Koslow, J. A., Brault, S., Dugas, J. and Page, F. (1985). Anatomy of an apparent year-class failure: the early life history of the 1983 Browns Bank haddock *Melanogrammus aeglefinus*. *Transactions of the American Fisheries Society* **114**, 478–489.

Kuhlmann, D. (1977). Laboratory studies on the feeding behaviour of the chaetognaths *Sagitta setosa* J. Muller and *S. elegans* Verril with special reference to fish eggs and larvae as food organisms. *Berichte der Deutschen Wissenschaftlichen Kommission für Meeresforschung* Bd. **25**, 163–171.

Lancraft, T. M. and Robison, B. H. (1980). Evidence of postcapture ingestion by midwater fishes in trawl nets. *Fishery Bulletin U.S.* **77**, 713–715.

Landry, M. R., Lehner-Fournier, J. M. and Fagerness, V. L. (1985). Predatory feeding behavior of the marine cyclopoid copepod *Corycaeus anglicus*. *Marine Biology* **85**, 163–169.

Larson, R. J. (1986). Seasonal changes in the standing stocks, growth rates, and production rates of gelatinous predators in Saanich Inlet, British Columbia. *Marine Ecology Progress Series* **33**, 89–98.

Lasker, R. (1975). Field criteria for survival of anchovy larvae: the relation between inshore chlorophyll maximum layers and successful first feeding. *Fishery Bulletin U.S.* **73**, 453–462.

Lasker, R. (1978). The relation between oceanographic conditions and larval anchovy food in the California Current: identification of factors contributing to recruitment failure. *Rapports et Procès-Verbaux des Réunions. Conseil International pour l'Exploration de la Mer* **173**, 212–230.

Laurence, G. C. (1982). Nutrition and trophodynamics of larval fish review, concepts, strategic recommendations and opinions. Fish Ecology III, *University of Miami Technical Report* **82008**, 123–147.

Laurence, G. C., Halavik, T. A., Burns, B. R. and Smigielski, A. S. (1979). An environmental chamber for monitoring "in situ" growth and survival of larval fishes. *Transactions of the American Fisheries Society* **108**, 197–203.

Leak, J. C. and Houde, E. D. (1987). Cohort growth and survival of bay anchovy

Anchoa mitchilli larvae in Biscayne Bay, Florida. *Marine Ecology Progress Series* **37**, 109–122.

Lebour, M. L. (1922). The food of planktonic organisms. *Journal of the Marine Biological Association of the United Kingdom* **12**, 644–677.

Lebour, M. L. (1923). The food of planktonic organisms II. *Journal of the Marine Biological Association of the United Kingdom* **13**, 70–92.

Lebour, M. L. (1925). Young anglers in captivity and some of their enemies. A study in a plunger jar. *Journal of the Marine Biological Association of the United Kingdom* **13**, 721–734.

Leggett, W. C. (1986). The dependence of fish larval survival on food and predator densities. *In* "The Role of Freshwater Outflows in Coastal Marine Ecosystems" (S. Skreslet, ed.), pp. 117–137. Springer-Verlag, Berlin.

Leis, J. M. (1982). Hawaiian Creediid fishes (*Crystallodytes cookei* and *Limnichthys donaldsoni*): development of eggs and larvae and use of pelagic eggs to trace coastal water movement. *Bulletin of Marine Science* **32**, 166–180.

Leong, R. and O'Connell, C. (1969). A laboratory study of particulate and filter feeding of northern anchovy (*Engraulis mordax*). *Journal of the Fisheries Research Board of Canada* **26**, 557–562.

Liem, K. F. (1984). Functional versatility, speciation, and niche overlap: are fishes different? *In* "Trophic Interactions within Aquatic Ecosystems" (D. G. Meyers and J. R. Strickler, eds.) pp. 269–305, American Association for the Advancement of Science, Washington, D.C.

Lillelund, K. (1967). Experimentelle Untersuchungen über den Einfluss carnivorer Cyclopiden auf der Sterblichkeit der Fischbrat. *Zeitschrift für Fischerei und deren Hilfwissenschaften* **15**, 29–43.

Lillelund, K. and Lasker, R. (1971). Laboratory studies of predation by marine copepods on fish larvae. *Fishery Bulletin U.S.* **69**, 655–667.

Lindstrom, T. (1955). On the fish size-food size. *Report Institute of Freshwater Research, Drottningholm* **36**, 133–147.

Lo, N. (1983). Re-estimation of three parameters associated with anchovy egg and larval abundance: temperature dependent incubation time, yolk-sac growth rate and egg and larval retention in mesh nets. *NOAA Technical Memorandum NMFS-SWFC* **31**, 1–33.

Lo, N. (1986) Modeling life-stage-specific instantaneous mortality rates, an application to northern anchovy, *Engraulis mordax*, eggs and larvae. *Fishery Bulletin U.S.* **84**, 395–407.

Lockwood, S. J. (1980). Density-dependent mortality in 0-group plaice (*Pleuronectes platessa* L.) populations. *Journal du Conseil. Conseil International pour l'Exploration de la Mer* **39**, 148–153.

Loftus, D. H. and Hulsman, P. F. (1986). Predation on larval lake whitefish (*Coregonus clupeaformis*) and lake herring (*C. artedii*) by adult rainbow smelt (*Osmerus mordax*). *Canadian Journal of Fisheries and Aquatic Science* **43**, 812–818.

Logachev, V. S. and Mordvinov, Y. E. (1979). Swimming speed and activity of larvae of round goby and some predatory crustaceans of the Black Sea. *Soviet Journal of Marine Biology* **5**, 227–229.

Lyons, J. and Magnuson, J. J. (1987). Effects of walleye predation on the population dynamics of small littoral-zone fishes in a northern Wisconsin lake. *Transactions of the American Fisheries Society* **116**, 29–39.

MacCall, A. D. (1980). The consequences of cannibalism in the stock-recruitment relationship of planktivorous pelagic fishes such as *Engraulis*. FAO, *International Oceanographic Commission Workshop Report* **28**, 201–220.

Margulies, D. (1986). Effects of food concentration and temperature on development, growth, survival and susceptibility to predation of larval white perch (*Morone americana*). Ph.D. Dissertation, University of Maryland, 168 pp.

Matsushita, K., Shimizu, M. and Nose, Y. (1982). Micro-distribution of anchovy eggs and larvae in Sagami Bay. *Bulletin of the Japanese Society of Scientific Fisheries* **48**, 355–362.

Mauchline, J. (1980). The biology of euphausiids. *Advances in Marine Biology* **18**, 373–595.

May, R. C. (1974). Larval mortality in marine fishes and the critical period concept. *In* "The Early Life History of Fish" (J.H.S. Blaxter, ed.), pp. 3–19, Springer-Verlag, Berlin.

May, R. M. (Ed.) (1984). Exploitation of marine communities: Report of the Dahlem Workshop in Exploitation of Marine Communities, Berlin 1984, April 1–6, Springer-Verlag, New York.

McGowan, J. A. and Miller, C. B. (1980). Larval fish and zooplankton community structure. *California Cooperative Oceanic Fisheries Investigation Reports* **21**, 29–36.

McGurk, M. D. (1986). Natural mortality of marine pelagic fish eggs and larvae: role of spatial patchiness. *Marine Ecology Progress Series* **34**, 227–242.

Mehl, S. and Westgard, T. (1983). The diet and consumption of mackerel in the North Sea. International Council for the Exploration of the Sea, C.M. 1983/H:34, 34 pp. (mimeo).

Mercer, M. (Ed.) (1982). Multispecies approaches to fisheries management advice. *Canadian Special Publication. Fisheries Aquatic Science* No. 59, 169 pp.

Mills, C. E. (1982). Patterns and mechanisms of vertical distribution of medusae and ctenophores. Ph.D. Dissertation. University of Victoria, Victoria, Canada, 384 pp.

Möller, H. (1980). Scyphomedusae as predators and food competitors of larval fish. *Meeresforschung* **28**, 90–100.

Möller, H. (1984). Reduction of a larval herring population by jellyfish predator. *Science* **224**, 621–622.

Monteleone, D. M. and Duguay, L. E. (1988). Laboratory studies of predation by the ctenophore *Mnemiopsis leidyi* on the early stages in the life history of the bay anchovy, *Anchoa mitchilli*. *Journal of Plankton Research* **10**, 359–372.

Murdoch, W. W. (1969). Switching in general predators: experiments on predator specificity and stability of prey populations. *Ecological Monographs* **39**, 335–354.

Murdoch, W. W. (1973). The functional response of predators. *Journal of Applied Ecology* **10**, 338–342.

Murdoch, W. W., Avery, S. and Smyth, M. E. B. (1975). Switching in predatory fish. *Ecology* **56**, 1094–1105.

Murphy, G. I. (1961). Oceanography and variations in the Pacific sardine population. *California Cooperative Fisheries Investigation Reports* **8**, 55–64.

Murphy, G. I. and Clutter, R. I. (1972). Sampling anchovy larvae with a plankton purse seine. *Fishery Bulletin U.S.* **70**, 789–798.

Nair, K. K. (1954). Medusae of the Tranvancore coast. Part II. Seasonal distribution. Kerala Univ., Trivandrum, India. Central Research Institute Bulletin, Ser. C, *Natural Science* **3**, 31–68.

Nicol, S. (1984). Cod end feeding by the euphausiid *Meganyctiphanes norvegica*. *Marine Biology* **80**, 29–33.

Nyberg, D. W. (1971). Prey capture in the largemouth bass. *American Midlands Naturalist* **86**, 128–144.

O'Brien, W. J. (1979). The predator-prey interaction of planktivorous fish and zooplankton. *American Scientist* **67**, 572–581.

O'Brien, W. J., Slade, N. A. and Vinyard, G. L. (1976). Apparent size as the determinant of prey selection by bluegill sunfish (*Lepomis macrochirus*). *Ecology* **57**, 1304–1310.

O'Brien, W. J., Evans, B. I. and Howick, G. L. (1986). A new view of the predation cycle of a planktivorous fish, white crappie (*Pomoxis annularis*). *Canadian Journal of Fisheries and Aquatic Sciences* **43**, 1894–1899.

O'Connell, C. P. (1980). Percentage of starving northern anchovy, *Engraulis mordax*, larvae in the sea as estimated by histological methods. *Fishery Bulletin U.S.* **78**, 475–489.

O'Connell, C. P. and Zweifel, J. (1972). A laboratory study of particulate and filter feeding of the Pacific mackerel *Scomber japonicus*. *Fishery Bulletin U.S.* **70**, 973–981.

Ohman, M. D. (1983). The effects of predation and resource limitation of the copepod *Pseudocalanus* sp. in Dabob Bay, a temperate fjord. Ph.D. Dissertation, University of Washington, Seattle, WA, 249 pp.

Ohman, M. D. (1986). Predator-limited population growth of the copepod *Pseudocalanus* sp. *Journal of Plankton Research* **8**, 673–713.

Øiestad, V. (1982). Application of enclosures to studies on the early life history of fishes. *In* "Marine Mesocosms" (M.R. Reeve, ed.), pp. 49–62. Springer-Verlag, New York.

Øiestad, V. (1985). Predation on fish larvae as a regulatory force, illustrated in mesocosm studies with large groups of larvae. *North Atlantic Fisheries Organization Council Studies* **8**, 25–32.

Olson, R. J. and Boggs, C. H. (1986). Apex predation by yellowfin tuna (*Thunnus albacarus*): independent estimates from gastric evacuation and stomach contents, bioenergetics, and cesium concentrations. *Canadian Journal of Fisheries and Aquatic Sciences* **43**, 1760–1775.

Olson, R. J. and Mullin, A. J. (1986). Recent developments for making gastric evacuation and daily ration determinations. *Environmental Biology of Fishes* **16**, 183–191.

Orians, G. H. and Janzen, D. H. (1974). Why are embryos so tasty? *American Naturalist* **108**, 581–592.

Outram, D. M. (1958). The magnitude of herring spawn losses due to bird predation on the west coast of Vancouver Island. *Fisheries Research Board of Canada Pacific Biological Station Progress Report* **111**, 9–13.

Palsson, W. A. (1984). Egg mortality upon natural and artificial substrata within Washington State spawning grounds of Pacific herring (*Clupea harengus pallasi*). M.S. Thesis, University of Washington, Seattle, WA, 191 pp.

Parrish, R. H., Nelson, C. S. and Bakun, A. (1981). Transport mechanisms and reproductive success of fishes in the California Current. *Biological Oceanography* **1**, 175–203.

Pauly, D. (1980). On the interrelationships between natural mortality, growth parameters, and mean environmental temperature in 175 fish stocks. *Journal du Conseil. Conseil International pour l'Exploration de la Mer* **39**, 175–192.

Pearre, S. (1980). Feeding by chaetognatha: the relation of prey size to predator size in several species. *Marine Ecology Progress Series* **3**, 125–134.

Pennington, M. (1985). Estimating the average food consumption by fish in the field from stomach contents data. *Dana* **5**, 81–86.

Pepin, P. (1987). Influence of alternative prey abundance on pelagic fish predation of larval fish: a model. *Canadian Journal of Fisheries and Aquatic Science* **44**, 222–227.

Pepin, P., Pearre, S. and Koslow, J. A. (1987). Predation on larval fish by Atlantic mackerel, *Scomber scombrus*, with a comparison of predation by zooplankton. *Canadian Journal of Fisheries and Aquatic Science* **44**, 2012–2018.

Persson, L. (1984). Food evacuation and models for multiple meals in fishes. *Environmental Biology of Fishes* **10**, 305–309.

Persson, L. (1986). Patterns of food evacuation of fishes: a critical review. *Environmental Biology of Fishes* **16**, 51–58.

Peterman, R. M. and Bradford, M. J. (1987). Wind speed and mortality rate of a marine fish, the northern anchovy (*Engraulis mordax*). *Science* **235**, 354–356.

Peterman, R. M. and Gatto, M. (1978). Estimation of functional responses of predators on juvenile salmon. *Journal of the Fisheries Research Board of Canada* **35**, 797–808.

Peterman, R. M., Bradford, M. J., Lo, N. C. and Methot, R. D. (1988). Contribution of early life stages to interannual variability in recruitment of northern anchovy (*Engraulis mordax*). *Canadian Journal of Fisheries and Aquatic Sciences* **45**, 8–16.

Peterson, I. and Wroblewski, J. S. (1984). Mortality rates of fishes in the pelagic ecosystem. *Canadian Journal of Fisheries and Aquatic Sciences* **41**, 1117–1120.

Pommeranz, T. (1981). Observations on the predation of herring (*Clupea harengus* L.) and sprat (*Sprattus sprattus* L.) on fish eggs and larvae in the southern North Sea. *Rapports et Procès-Verbaux des Réunions. Conseil International pour l'Exploration de la Mer* **178**, 402–404.

Purcell, J. E. (1980). Influence of siphonophore behavior upon their natural diets: evidence for aggressive mimicry. *Science* **209**, 1045–1047.

Purcell, J. E. (1981). Feeding ecology of *Rhizophysa eysenhardti*, a siphonophore predator of fish larvae. *Limnology and Oceanography* **26**, 424–432.

Purcell, J. E. (1983) Digestion rates and assimilation efficiencies of siphonophores fed zooplankton prey. *Marine Biology* **73**, 257–261.

Purcell, J. E. (1984). Predation on fish larvae by *Physalia physalis*, the Portuguese man of war. *Marine Ecology Progress Series* **19**, 189–191.

Purcell, J. E. (1985). Predation on fish eggs and larvae by pelagic cnidarians and ctenophores. *Bulletin of Marine Science* **37**, 739–755.

Purcell, J. E. (1986). Jellyfish as predators of larval herring at spawning grounds in British Columbia. *Canadian Manuscript Report Fisheries and Aquatic Science* **1871**, 139–140.

Purcell, J. E. (Unpublished). Predation on fish larvae and eggs by the hydromedusa *Aequorea victoria* at a herring spawning ground in British Columbia, 1983.

Purcell, J. E., Siferd, T. S. and Marliave, J. B. (1987). Vulnerability of larval herring (*Clupea harengus pallasi*) to capture by the jellyfish *Aequorea victoria*. *Marine Biology* **94**, 157–162.

Ricker, W. E. (1954). Stock and recruitment. *Journal of the Fisheries Research Board of Canada* **11**, 559–623.

Riessen, H. P., O'Brien, W. J. and Loveless, B. (1984). An analysis of the components of *Chaoborous* predation on zooplankton and the calculation of relative prey vulnerabilities. *Ecology* **65**, 514–522.

Robertson, D. A. (1981). Possible functions of surface structure and size in some planktonic eggs of marine fishes. *New Zealand Journal of Marine and Freshwater Research* **15**, 147–153.

Rothschild, B. J. (1986). Dynamics of marine fish populations. Harvard University Press, Cambridge, Massachusetts, 277 pp.

Rothschild, B. J. and Rooth, C. G. H. (1982). Fish Ecology III. *University of Miami Technical Report* No. 82008.

Runge, J. A. (1981). Egg production of *Calanus pacificus* Brodsky and its relation to seasonal changes in phytoplankton availability. Ph.D. Dissertation, University of Washington, 116 pp.

SCOR (1981). Draft Report Scientific Committee on Ocean Research (SCOR) and Advisory Committee of Experts on Marine Resources (ACMRR). Working group # 67. Oceanography, Marine Ecology and Living Resources, 17 pp.

Seghers, B. H. (1974). Role of gill rakers in size-selective predation by lake whitefish, *Coregonus clupeaformis* (Mitchill). *Verhandlungen Internationale Vereinigung für Limnologie Theoretische und Angewandte* **19**, 2401–2405.

Selgeby, J. H., MacCallum, W. R. and Swedberg, D. V. (1978). Predation by rainbow smelt (*Osmerus mordax*) on lake herring (*Coregonus artedii*) in western Lake Superior. *Journal of the Fisheries Research Board of Canada* **35**, 1457–1463.

Shapiro, D. Y., Hensley, D. A., Appeldoorn, R. S. (1988). Pelagic spawning and egg transport in coral-reef fishes: a skeptical overview. *Environmental Biology of Fishes* **22**, 3–14.

Sharp, G. D. (1980). Workshop on the effects of environmental variation on the survival of larval pelagic fishes. *Intergovernmental Oceanographic Commission Workshop Report* No. 28, FAO, Rome, 323 pp.

Sharp, G. D. and Csirke, J. (1983). Proceedings of the expert consultation to examine changes in abundance and species composition of neritic fish resources. *FAO Fisheries Report* No. 291, 1224 pp.

Sheader, M. and Evans, F. (1975). Feeding and gut structure of *Parathemisto gaudichaudi* (Guerin) (Amphipoda, Hyperiidea). *Journal of the Marine Biological Association of the United Kingdom* **55**, 641–656.

Shepherd, J. G. and Cushing, D. H. (1980). A mechanism for density-dependent survival of larval fish as the basis of a stock-recruitment relationship. *Journal du Conseil. Conseil International pour l'Exploration de la Mer* **39**, 160–167.

Shepherd, J. G., Pope, J. G. and Cousens, R. D. (1984). Variations in fish stocks and hypotheses concerning their links with climate. *Rapports et Procès-Verbaux des Réunions. Conseil International pour l'Exploration de la Mer* **185**, 255–267.

Sinclair, M. and Tremblay, M.J. (1984). Timing of spawning of Atlantic herring (*Clupea harengus harengus*) populations and the match-mismatch theory. *Canadian Journal of Fisheries and Aquatic Sciences* **41**, 1055–1065.

Sissenwine, M. P. (1984). Why do fish populations vary? In "Exploitation of Marine Communities" (R. May, ed.), pp. 59–94, Springer-Verlag, Berlin.

Skud, B. E. (1982). Dominance in fishes: the relation between environment and abundance. *Science* **216**, 144–149.

Smith, P. E. (1981). Fisheries on coastal pelagic schooling fish. In "Marine Fish Larvae: Morphology, Ecology and Relation to Fisheries" (R. Lasker, ed.), pp. 1–31, Washington SeaGrant, University of Washington Press, Seattle, WA.

Smith, P. E. (1985). Year class strength and survival of 0-group clupeoids. *Canadian Journal of Fisheries and Aquatic Sciences* **42**, 69–82.

Smith, R. E. and Kernehan, R. J. (1981). Predation by the free-living copepod,

Cyclops bicuspidatus thomasi, on larvae of the striped bass and white perch. *Estuaries* **4**, 81–83.

Solomon, M. E. (1949). The natural control of animal populations. *Journal of Animal Ecology* **18**, 1–35.

Stedman, R. M. and Argyle, R. L. (1985). Rainbow smelt (*Osmerus mordax*) as predators on young bloaters (*Coregonus hoyi*) in Lake Michigan. *Journal of Great Lakes Research* **11**, 40–42.

Steinfeld, J. D. (1972). Distribution of Pacific herring spawn in Yaquina Bay, Oregon, and observations of mortality through hatching. M.S. Thesis. Oregon State University, Corvallis, Oregon, 75 pp.

Stevenson, J. C. (1962). Distribution and survival of herring larvae (*Clupea pallasi* Valenciennes) in British Columbia waters. *Journal of Fisheries Research Board of Canada* **19**, 735–810.

Swanberg, N. (1974). The feeding behavior of *Beroe ovata*. *Marine Biology* **24**, 69–76.

Tarby, M. J. (1974). Characteristics of yellow perch cannibalism in Oneida Lake and the relation to first year survival. *Transactions of the American Fisheries Society* **103**, 462–471.

Theilacker, G. H. (1986). Starvation-induced mortality of young sea-caught jack mackerel, *Trachurus symmetricus*, determined with histological and morphological methods. *Fishery Bulletin U.S.* **84**, 1–17.

Theilacker, G. H. (1988) Euphausiid predation on larval anchovy at two contrasting sites off California determined with an elispot immunoassay. In "Immunochemical Approaches to Coastal, Estuarine and Oceanographic Questions" (C. M. Yentsch, F. C. Mague and P. K. Horan, eds), pp. 304–311. Springer-Verlag, New York.

Theilacker, G. H. and Dorsey, K. (1980). Larval fish diversity, a summary of laboratory and field research. *Intergovernmental Oceanographic Commission Workshop Report* **28**, 105–142.

Theilacker, G. H. and Lasker, R. (1974). Laboratory studies of predation by euphausiid shrimps on fish larvae. *In* "The Early Life History of Fish" (J.H.S. Blaxter, ed.), pp. 287–300, Springer-Verlag, Berlin.

Theilacker, G. H., Kimbrell, A. S. and Trimmer, J. S. (1986). Use of an ELISPOT immunoassay to detect euphausiid predation on anchovy larvae. *Marine Ecology Progress Series* **30**, 127–131.

Tinbergen, L. (1960). The dynamics of insect and bird populations in pinewoods. *Archives Neerlandaises de Zoologie* **13**, 259–473.

Turner, J. T., Tester, P. A. and Hettler, W. F. (1985). Zooplankton feeding ecology. A laboratory study of predation on fish eggs and larvae by the copepods *Anomalocera ornata* and *Centropages typicus*. *Marine Biology* **90**, 1–8.

Valdés, E. S., Shelton, P. S., Armstrong, M. J. and Field, J. G. (1987). Cannibalism in South African anchovy: egg mortality and egg consumption rates. *South African Journal of Marine Science* **5**, 613–622.

Veer, H. W. van der (1985). Impact of coelenterate predation on larval plaice *Pleuronectes platessa* and flounder *Platichthys flesus* stocks in the western Wadden Sea. *Marine Ecology Progress Series* **25**, 229–238.

Veer, H. W. van der (1986). Immigration, settlement, and density-dependent mortality of a larval and early postlarval plaice (*Pleuronectes platessa*) population in the western Wadden Sea. *Marine Ecology Progress Series* **29**, 223–236.

Veer, H. W. van der and Bergman, M. J. N. (1987). Predation by crustaceans on a newly settled 0-group plaice *Pleuronectes platessa* population in the western Wadden Sea. *Marine Ecology Progress Series* **35**, 203–215.

Veer, H. W. van der and Oorthuysen, W. (1985). Abundance, growth and food demand of the scyphomedusa *Aurelia aurita* in the western Wadden Sea. *Netherlands Journal of Sea Research* **19**, 38–44.

Veer, H. W. van der and Zijlstra, J. J. (1982). Predation of flatfish larvae by *Pleurobrachia pileus* in coastal waters. ICES, C.M. 1982/G: 16.

Veer, H. W. van der, Garderen, H. van and Zijlstra, J. J. (1983). Impact of coelenterate predation on larval fish stocks in the coastal zone of the southern North Sea. International Council for the Exploration of the Sea, C.M. 1983/L: 8 (mimeo).

Vinyard, G. L. (1980). Differential prey vulnerability and predator selectivity: effects of evasive prey on bluegill (*Lepomis macrochirus*) and pumpkinseed (*L. gibbosus*) predation. *Canadian Journal of Fisheries and Aquatic Science* **37**, 2294–2299.

Waddy, S. L. and Aiken, D. E. (1985). Immunofluorescent localization of American lobster egg yolk protein in the alimentary tract of the nemertean *Pseudocarcinonemertes homari*. *Canadian Journal of Fisheries and Aquatic Sciences* **42**, 357–359.

Walter, C. B., O'Neill, E. O. and Kirby, R. (1986). "Elisa" as an aid in the identification of fish and molluscan prey of birds in marine ecosystems. *Journal of Experimental Marine Biology and Ecology* **96**, 97–102.

Ware, D. M. (1971). Predation by rainbow trout (*Salmo gairdneri*): the effect of experience. *Journal of the Fisheries Research Board of Canada* **28**, 1847–1852.

Ware, D. M. (1975). Relation between egg size, growth and natural mortality of larval fish. *Journal of the Fisheries Research Board of Canada* **32**, 2503–2512.

Webb, P. W. (1981). Responses of northern anchovy, *Engraulis mordax*, larvae to predation by a biting planktivore, *Amphiprion percula*. *Fishery Bulletin U.S.* **79**, 727–735.

Webb, P. W. (1986). Effect of body form and response threshold on the vulnerability of four species of teleost prey attacked by largemouth bass (*Micropterus salmoides*). *Canadian Journal of Fisheries and Aquatic Science* **43**, 763–771.

Welch, D. W. (1986). Identifying the stock-recruitment relationship for age-structured populations using time-invariant matched linear filters. *Canadian Journal of Fisheries and Aquatic Science* **43**, 108–123.

Werner, E. E. and Gilliam, J. F. (1984). The ontogenetic niche and species interactions in size-structured populations. *Annual Review of Ecology and Systematics* **15**, 393–425.

Westernhagen H. von (1976). Some aspects of the biology of the hyperiid amphipod *Hyperoche medusarum*. *Helgoländer Wissenschaftliche Meeresuntersuchungen* **28**, 43–50.

Westernhagen, H. von and Rosenthal, H. (1976). Predator-prey relationship between Pacific herring, *Clupea harengus* Pallasi, larvae and a predatory hyperiid amphipod, *Hyperoche medusarum*. *Fishery Bulletin U.S.* **74**, 669–674.

Westernhagen, H. von, Rosenthal, H., Kerr, S. and Fürstenberg, G. (1979). Factors influencing predation of *Hyperoche medusarum* (Hyperiida: Amphipoda) on larvae of the Pacific herring *Clupea harengus pallasi*. *Marine Biology* **51**, 195–201.

Wood, C. C. (1987). Predation of juvenile Pacific salmon by the common merganser (*Mergus merganser*) on eastern Vancouver Island. I: Predation during the seaward migration. *Canadian Journal of Fisheries and Aquatic Science* **44**, 941–949.

Yamashita, Y., Aoyama, T. and Kitagawa, D. (1984). Laboratory studies of predation by the hyperiid amphipod *Parathemisto japonica* on larvae of the

Japanese sand-eel *Ammodytes personatus*. *Bulletin of the Japanese Society of Scientific Fisheries* **50**, 1089–1093.

Yamashita, Y., Kitagawa, D. and Aoyama, T. (1985). A field study of predation of the hyperiid amphipod *Parathemisto japonica* on larvae of the Japanese sand eel *Ammodytes personatus*. *Bulletin of the Japanese Society of Scientific Fisheries* **51**, 1599–1607.

Yen, J. (1982). Predatory feeding ecology of *Euchaeta elongata* Esterly, a marine planktonic copepod. Ph.D. Dissertation. University of Washington, Seattle, WA, 130 pp.

Yen, J. (1983). Effects of prey concentration, prey size, predator life stage, predator and season on predation rates of the carnivorous copepod *Euchaeta elongata*. *Marine Biology* **75**, 69–78.

Yen, J. (1987). Predation by carnivorous marine copepod, *Euchaeta norvegica* Boeck, on eggs and larvae of the Norwegian cod, *Gadus morhua* L. *Journal of Experimental Marine Biology and Ecology* **112**, 283–296.

Yin, M. C. and Blaxter, J. H. S. (1987). Escape speeds of marine fish larvae during early development and starvation. *Marine Biology* **96**, 459–468.

Zaret, T. M. (1980). Predation and freshwater communities. Yale University Press, New Haven, Connecticut, 187 pp.

Zaret, T. M. and Kerfoot, W. C. (1975). Fish predation on *Bosmina longirostris*: body-size selection versus visibility selection. *Ecology* **56**, 232–237.

Zottoli, S. J. and Horne, C. van (1983). Posterior lateral line afferent and efferent pathways within the central nervous system of the goldfish with special reference to the Mauthner Cell. *Journal of Comparative Neurology* **219**, 100–111.

Recent Studies on Spawning, Embryonic Development, and Hatching in the Cephalopoda

S. v. Boletzky

Laboratoire Arago, C.N.R.S. and University of Paris 6 (U.R.A. 117),
F-66650 Banyuls-sur-Mer

ADVANCES IN MARINE BIOLOGY,
VOLUME 25 ISBN 0–12–026125–1

I. Introduction

Cephalopods are exclusively marine molluscs and should be included among the organisms that are of general interest to marine biologists. To give an example, the index of "The Biology of Marine Animals" (Nicol, 1967) contains about 130 page references for various cephalopod species, which are considered under numerous aspects throughout the book. By contrast, one finds virtually nothing on early ontogenetic processes, especially embryonic development. This is indeed considered by most people to lie outside the scope of marine biology. For embryological information, one has to use the special literature that is now generally classified under Developmental Biology. In practice, this is not so easy, if one desires a reasonably complete, up-to-date overview of all the studies dealing with cephalopod embryos. The last comprehensive coverage appeared a decade ago, with a bibliography up to and including 1975, plus a few references in 1976 (Fioroni, 1978). About one hundred papers dealing with cephalopod eggs and embryos have appeared since this review. The present article provides a brief overview of these recent studies placing them in the chronological sequence of embryogenesis. Studies covering early stages of embryonic development as well as later ones will be cited at least once in the section dealing with the earliest stage considered. Most of these investigations ultimately rest on the basic work by Naef (1928).

Of the roughly 700 species of extant cephalopods, only a small number is known in terms of the whole life cycle, and among these two or three dozen species, only a few have been the subject of intensive embryological investigation (cf. Boyle, 1983, 1987). A step forward in providing additional data might be taken each time a seagoing or beach-combing biologist finds eggs and identifies them as belonging to a cephalopod. The likelihood that such eggs represent something new to science is greater in offshore waters than inshore, but even the shore line can still provide surprising "finds", especially in areas that have not yet been studied systematically. The way to deal with such material is indicated by Roper and Sweeney (1983). A review of cephalopod maintenance and rearing facilities, with emphasis on water quality and food choice, is given by Boletzky and Hanlon (1983). Bottom-living cephalopods, collected alive and transferred to an aquarium providing appropriate temperature and salinity conditions, can often be maintained in captivity for long periods. The females may then lay fertile eggs, thus allowing a complete coverage of the embryonic development.

This is not the place to go into details of evolutionary background of cephalopod development; the subject is covered by Fioroni (1977, 1979a,b, 1980, 1982a,b,c,d, 1983) and by Boletzky (1988a,b). Here it will be sufficient to say that cephalopods are characterized by a mode of blastulation (cleavage) that is unique within the phylum Mollusca, whereas the gastrula exhibits the basic lay-out of organ areas according to the molluscan bauplan. Cephalopods have developed a sophisticated buoyancy apparatus from a simple molluscan shell (Bandel and Boletzky, 1979). They have highly developed sense organs; the eyes and the statocysts are surprisingly similar to the analogous organs of vertebrates (cf. O'Dor and Webber, 1986). Only one representative genus of the primitive ectocochleate chambered shell design has survived to the present time; this is the Pearly Nautilus with a half dozen species living in the Indo-Pacific. A highly modified chambered shell exists in the endocochlean (coleoid) *Spirula* and sepiids (*Sepia, Sepietta, Hemisepius*). Whereas the coiled shell of *Spirula* contains wide chambers, the "cuttlebone" of sepiids retains only the flattened lamellae representing the "septal necks" of typical chambered shells. Both types of inner shells, the endogastrically coiled *Spirula* shell and the nearly straight (only slightly endogastrically curved) cuttlebone strongly contrast with the exogastric coiling of the external shell of *Nautilus*. All the other forms of coleoid squids and octopods have typically reduced, uncalcified inner shells or shell relics. The integument of both the mantle and the head-foot contains a complex system of pigment cells and reflecting elements allowing extremely rapid changes of all-over aspect due to the nerve-controlled expansion or contraction of the chromatophore cells, which may be combined with other muscular contractions in the skin. This system, especially the chromatophores with their muscular tenters, is unique in the animal kingdom.

II. Eggs and Egg Masses

Cephalopods are strict gonochorists (dioeceous). The gonads and accessory sexual organs are differentiated at different times in early development. External sexual dimorphism becomes distinct only at sub-adult stages.

A description of embryonic development would be incomplete if no mention were made of the envelopes or capsules in which the ova are laid. The first envelope produced is the chorion which is secreted by the follicle during late oogenesis. Mature ova, which have broken free of the follicular complex, are released through the oviduct(s), either one by one, or in successive series of several to many eggs at a time (see Boletzky, 1986). The former mode is characteristic (as far as is known) of the decapodan order Sepioidea, i.e. cuttlefishes such as *Sepia* (Okutani, 1978; Natsukari, 1979; Boletzky, 1983a), sepiolid squids (Boletzky, 1983b; Singley, 1983; Bergstrom and Summers, 1983) and the Pygmy Cuttlefish *Idiosepius* (Natsukari, 1970; for complete systematic lists, see Voss, 1977; Clarke, 1988). Unfortunately we know nothing of egg laying in *Spirula*.

In the second decapodan order Teuthoidea ("squids"), laying of single eggs represents a special mode (Okiyama and Kasahara, 1975); typically eggs are released in series (Misaki and Okutani, 1976; Natsukari, 1976; Suzuki *et al.*, 1979; Hixon, 1983; Summers, 1983; Worms, 1983; O'Dor, 1983; Okutani, 1983; Sabirov *et al.*, 1987; Segawa, 1987; Boletzky, 1987c).

In both the Sepioidea and the Teuthoidea, each egg cell with its chorion becomes surrounded by some jelly during egg laying. An inner layer is provided by the oviducal gland; an outer layer is added thereafter by the paired nidamental glands which lie in the mantle cavity. Except in ommastrephid squids (see below), this nidamental jelly is spirally wrapped around the eggs. Thus in Sepioidea, each egg is individually envelopped, whereas in the majority of the Teuthoidea, a series of eggs is contained within a sheet of nidamental jelly. In the ommastrephid squids, the nidamental jelly forms an internally amorphous, globular mass (O'Dor *et al.*, 1982). Another interesting modification exists in a group of pelagic squids (Enoploteuthinae of the oegopsid family Enoploteuthidae) which lack nidamental glands. In these animals, the oviducal glands are very large and have an aspect very similar to nidamental glands; the jelly produced by them is secreted around individual eggs that are probably released separately (Young and Harman, 1985; Young *et al.*, 1985).

The monotypic order Vampyromorpha (represented by the pelagic *Vampyroteuthis infernalis*) shows a situation somewhat similar to that described above, in that independent nidamental glands are lacking, whereas the oviducal glands are well-developed (though less voluminous than those of the Enoploteuthinae). Unfortunately no new data on the eggs have become available since the work of Pickford (1949). The order Octopoda is also characterized by the absence of nidamental glands. In the sub-order Cirrata (the finned octopods, most of which are abyssal bentho-pelagic animals), only one oviduct is differentiated; it has a very large oviducal gland which produces the capsule material for each of the very large eggs (Boletzky, 1982a) (Fig. 1). In the sub-order Incirrata, each paired oviduct also has a well-developed oviducal gland, but the secretions of these glands are not used to encapsulate the entire egg (Boletzky, 1978–79). The form of the incirrate egg is unique in that the chorion is drawn out into a stalk (at the side opposite to the micropyle). The oviducal gland secretions are used only to surround the end of the chorion stalk, which is either cemented directly to a hard substratum, or to other chorion stalks, so that a festoon-like egg string is formed with a common axis. Egg strings are then either cemented to a substratum, or are carried around by the female. The common feature of all Incirrata is that the female actively protects her eggs until the young hatch out. In the pelagic incirrates, there is a wide variety of brooding modes, all being achieved by the actions of the arms. One extreme situation exists in

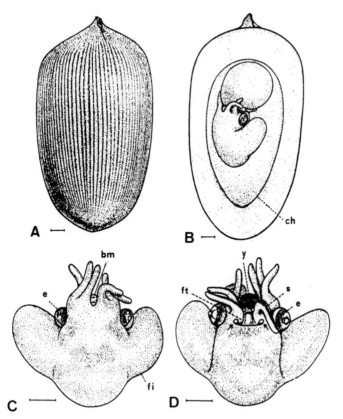

Fig. 1. An egg of a cirrate octopod with an embryo at a late developmental stage. (A) The egg, showing the regular longitudinal ribs of the outer case. (B) Egg shell opened to expose the much smaller chorion (ch) surrounded by gelatinous material. The embryo is not tightly enclosed in the chorion; there is a large volume of perivitelline fluid (in contrast to what has been seen in other cirrate eggs). (C) Dorsal view of the embryo (without the outer yolk sac), showing the buccal mass (bm) between the bases of the dorsal arms, the eyeballs (e) not yet covered by the corneal skin, and the large fins (fi). (D) Ventral view of the embryo showing the yolk (y) sectioned, the comparatively long arms with a row of sucker (s) rudiments, the eyeballs (e) with the eye lens, the short funnel tube (ft) emerging from the mantle aperture, the lateral ends of which are marked by the olfactory vesicles (arrows). Scale bar = 1 mm (from Boletzky, 1982a).

Argonauta, in which glandular cells of the dorsal arms secrete material to form a calcified brood shell; the other extreme case of modification is the ovovivipary of *Ocythoe*. In the bottom-living incirrate family Octopodidae, the eggs are generally cemented (either individually or in strings) to a hard substratum, but in the sub-family Octopodinae, several species are known in which the female carries the egg strings in a fashion reminiscent of the pelagic incirrates (Forsythe and Hanlon, 1985; also see Boletzky, 1984).

In *Nautilus*, the eggs are laid one by one. The chorion is surrounded by some gelatinous material and an outer, nidamental envelope which forms a double layer except at the apex (cf. Arnold and Carlson, 1986; Arnold, 1987). The space between the inner and the outer layer of this tough envelope is open to the surrounding sea water via a series of channels formed by the sub-apical folds of the outer layer.

The various forms of encapsulation in cephalopods must of course allow two processes:

(1) the passage of spermatozoids to the micropyle, and
(2) the final hatching of the young animals at the close of embryonic development (Boletzky, 1986).

All the freshly formed material enveloping the eggs is very soft and gelatinous for at least a few hours (outer "shells" being formed by subsequent hardening). The structural modifications occurring during embryonic development are still poorly known, especially with regard to the contribution of the developing embryo in terms of the biochemistry of the whole capsule complex (cf. Gomi *et al.*, 1986; Boletzky, 1987c).

III. Fertilization

A typical feature of cephalopods is the production of spermatophores by a complex of glands of the male duct. These sperm carriers allow the mass transfer of spermatozoids to the female, where they may be stored for some time before the sperm is used in portions during spawning. In the octopods, spermatophores are introduced into the oviduct, and here fertilization is truly internal (Mangold, 1987). This, of course, is the prerequisite of ovovivipary as in the pelagic *Ocythoe*. An intermediate situation exists between this extreme and the generalized synchronization of spawning and fertilization of the eggs in *Argonauta* and the closely related *Tremoctopus*, where the eggs go through early cleavage stages still inside the oviduct (cf. Boletzky and Centelles, 1978–79).

In the decapods (Sepioidea and Teuthoidea), the spermatophores are generally deposited in some sort of receptacle, which lies either in the mantle cavity or under the buccal mass, or is "produced" by actions of the male (see Mangold, 1987). Thus spermatozoa have to cross at least some of the freshly produced jelly to arrive at the micropyle of the egg.

Artificial fertilization, in the absence of the gelatinous capsule material, is possible in decapod eggs. Subsequent problems related to insufficient swelling of the chorion, resulting from the absence of gelatinous coats, can be overcome by wrapping artificially fertilized eggs with jelly from freshly laid egg masses (Arnold, 1984a).

IV. Cleavage

The phase of cleavage or blastulation comprises the developmental stages 0 to I as defined by Naef (1928), or stages 1 to 9 as defined by Arnold (1965), Lemaire (1970), Yamamoto (1982) and Segawa (1987).

In contrast to other molluscan eggs, the cleavage in cephalopods is partial and discoidal; it involves only the cytoplasmic cap at the animal pole of the zygote (Kao, 1985). Cleavage furrows cut across this blastodisc, but do not penetrate into the yolk mass. In small eggs, the outer ends of the cleavage furrows approach, but never actually reach the equator of the egg. In the larger eggs the furrow length represents only a small fraction of the egg diameter, so that the area of cleavage forms a flat disc rather than a hollow cap.

The general geometry of blastulation has been described by many authors since the mid-nineteenth century (see Naef, 1928; Arnold, 1976; Marthy, 1976; Singley, 1977). The first cleavage furrow has been shown to mark the future symmetry plane of the embryo; it appears near the polar bodies that have been extruded after fertilization (Crawford, 1985). The second cleavage furrow lies roughly perpendicular to the first, a short distance from the polar bodies. In decapod eggs, the base of the second furrow does not form a right angle with the first furrow (Fig. 2). Two of the four segments are larger, and these are subsequently recognized as marking the anterior (buccal) end of the embryo, whereas the smaller segments belong to the posterior end. The third cleavage furrow is again roughly perpendicular to the second, but not exactly at right angles with the second furrow. The resulting eight segments thus show a pseudo-radial arrangement. In the decapod eggs, the slanting position of the second furrow entails a peculiar situation in the posterior segments, where the third furrow comes to lie close to and parallel with the first cleavage furrow. The fourth cleavage step, with furrow orientation roughly parallel to the second furrow, cuts off four small blastomeres ("micromeres") in the centre of the cleavage territory, one from each of the admedian segments, while the outer segments are again halved. This is the beginning of the cellular formation in the blastodisc, which proceeds from the centre to the periphery during the subsequent cleavage stages. In spite of the "deformed" angles between new furrows and the respective preceding ones, one can make out an orderly tilt pattern in the derivation of cleavage furrows (or mitotic spindles, which lie roughly perpendicular to the resulting

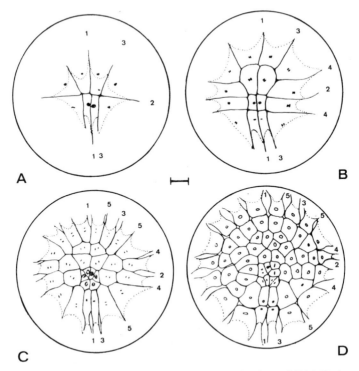

FIG. 2. Early cleavage in *Loligo pealei* (camera lucida drawings of FAA-Feulgen treated embryos). Numerals indicate the cleavage number of the corresponding furrow. Dotted lines indicate the positions of undercutting furrow bases. Nuclei are drawn as they appear with Feulgen stain. Scale bar = 0.1 mm.
(A) Third cleavage (Stage 6 of Arnold). The third cleavage furrows are unequal in the posterior half and symmetrical in the right and left halves.
(B) Fourth cleavage (Stage 7). Fourth cleavage is roughly parallel to the second furrow and forms the first four blastomeres.
(C) Fifth cleavage (Stage 8). Cleavage planes are radial except in the blastocones adjacent to the second furrow and in the posterior-medial blastocones.
(D) Sixth cleavage (Stage 9). All blastocones and the large blastomeres abutting the second furrow in Stage 8 have cleaved trans-radially. The orientation in central blastomeres varies (from Singley, 1977).

furrow). Starting with an orientation that could be called X for first cleavage, Y for second cleavage, again X for third cleavage, and so on, it turns out that the normal sequence X–Y–X is changed to X–X–Y in certain blastomere derivations (cf. Boletzky, 1988a).

Arnold (1976) studied the mechanism of cleavage furrow formation and found that the active element is formed by a band of 50 Å filaments which are linked to the surface membrane. Due to the convex curvature of the blastodisc surface, contraction of these filaments results in the formation of a furrow. Microsurgical experiments confirmed the impression derived from simple observation of the progress of cleavage under the microscope, that furrows are always anchored to previously formed furrows. Arnold also studied the effect of cytochalasin B, a drug which prevents actin polymerization, on the cleavage in *Loligo pealei* eggs; he found that, depending on dosage and length of application, existing furrows irreversibly disappeared.

Singley (1977) documented cleavage stages in *L. pealei* and found that at the stage preceding the onset of gastrulation, there are about 230 blastomeres; they are surrounded by 24 or 26 so-called blastocones, ray-like peripheral elements that remain in syncytial continuity with the uncleaved egg cortex. These elements have been described in the classic studies (e.g. Naef, 1928) on the basis of histological analysis. They are now better known in terms of their ultrastructural relations to the neighbouring blastomeres and to the underlying yolk thanks to the electron microscopical studies of Arnold and Singley.

V. Gastrulation

From the observations made by Marthy (1976) on *Loligo vulgaris* and by Singley (1977) on *L. pealei*, it is clear that the onset of gastrulation is marked by a shearing movement, achieved by two cell populations in relation to one another: the outer ring of blastomeres (two rows wide) lying next to the blastocones undergoes a "subduction" under the blastomeres forming the central blastodisc. Thus two cell layers are formed. The newly formed "inner layer" continues to grow towards the central area of the disc and forms the mesendoderm. Below it the nuclei of the former blastocones divide further to form a continuous yolk syncytium. Meanwhile the "central disc", which forms the ectoderm, becomes enlarged and grows beyond the periphery of the mesendoderm. The peripheral part (representing the blastopore lip) grows over the yolk mass to the opposite egg pole. Together with some scattered mesodermal elements derived from the inner layer of the embryo proper (Marthy, 1985), it thus forms a closed outer yolk sac (see also Segmüller and Marthy, 1984).

This process is essentially identical in all cephalopods. However, variations exist in the timing of the onset of gastrulation (which may start immediately after the seventh cleavage stage or somewhat later, according to species-specific egg size) and in the length of the whole process of yolk sac formation. In small eggs, the blastular and early gastrular cap is relatively large, so that the free surface of the uncleaved yolk mass is rather rapidly covered (O'Dor *et al.*, 1982; Boletzky, 1988a,b). By contrast, this surface is extremely large compared to the embryo proper in larger eggs. The speed of cleavage, as that of subsequent embryonic processes, is strictly temperature-dependent.

VI. Organogenesis

The end of gastrulation proceeds quite smoothly into the morphogenetic processes leading to the differentiation of organ areas within which distinct organ rudiments will finally arise. In small eggs, this gradual process is less clearly observable than in larger ones, so that the impression may arise that organogenesis only starts when the greater part of the yolk mass is covered by the germinal disc.

In larger eggs, where the formation of the outer yolk sac envelope takes a relatively longer time, it is evident from direct observations that organ areas are differentiated close to the animal pole before the edge of the gastrular cap has reached the equator of the yolk mass. With the continuing growth of the yolk sac envelope, the embryo proper also increases in size. The studies of Arnold and Williams-Arnold (1976) and Marthy (1976) suggest that some organ-determining information is stored in the cortex of the ovum, in a ring-shaped zone within the animal hemisphere. This zone would correspond to the blastocones and adjacent blastomeres that become covered by the ectoderm during gastrulation (cf. Boletzky, 1988a).

Cartwright and Arnold (1980, 1981) analysed the intercellular bridges within the blastoderm at these early stages. They suggest that gene products are shared and protein synthesis of groups of bridged cells is synchronized.

As soon as the outer yolk sac envelope is closed at the vegetal pole, the cap forming the embryo proper (where organ rudiments are now fairly distinct) undergoes a progressive radial contraction accompanied by several folding events: formation of primary eye folds (optic vesicle), invagination of the statocysts, stomodaeal invagination, and shell sac formation by the closure of a circular fold on the mantle surface (Boletzky, 1987a, 1988a,b). By a subsequent process of micro-invagination, three sets of small ectodermic vesicles are formed in the cephalopodium (see Wittland and Fioroni, 1982); their function is still unknown (Fig. 3).

A. *Circulatory System*

The concentration and elevation of embryonic tissue masses is accompanied by the formation of blood spaces that are in continuity with the blood lacuna surrounding the yolk mass inside the outer sac (Boletzky, 1987b). This primary circulatory system is flushed by the regular waves of contraction travelling over the surface of the yolk sac envelope. The blood sinuses lying inside the embryo proper become differentiated as part of the definitive venous system. With the subsequent formation of the arterial system and its connection to the primitive venous system, the function of the outer yolk sac

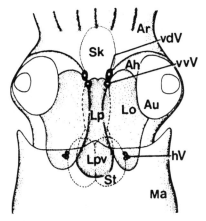

Fig. 3. Reconstruction of the cephalic region of a *Sepia* hatchling showing the positions of ectoderm-derived vesicles, one pair of anterior dorsal (vdV) and one pair of anterior ventral vesicles (vvV) lying behind the buccal mass (Sk), which is surrounded by the bases of the arms (Ar). The brain is drawn stippled, with the optic lobes (Lo) lying next to the eye (Au) and the orbital cavity (Ah), the pedal lobe (Lp) and the palliovisceral lobe (Lpv), which partly overlies the statocysts (St). Here lie the posterior vesicles (hV). Ma = mantle (From Wittland and Fioroni, 1982).

as a pump is gradually replaced by the systemic heart and the branchial hearts in later organogenesis.

It must be stressed here that the outer yolk sac is a complex embryonic "organ" that functions both as a respiratory and circulatory support system with a musculature which is active long before the embryo proper has formed functional gills and hearts. In addition to these functions, the yolk sac envelope also acts as a stirrer; the entire surface is covered with cilia that achieve a continuous circulation of the perivitelline fluid around the embryo. An exception are the embryos of ommastrephid squids, in which the outer yolk sac is extremely reduced in size (O'Dor *et al.*, 1982; Boletzky, 1988a,b).

A peculiar type of cell ("vacuolized round cells") which appears to occur only in the branchial hearts and in the ophthalmic sinus of the embryo has been described by Meister (1977). More recently the same author studied the ultrastructure of these cells, which are thought to be involved in the cellular defence system (Sundermann, 1980).

B. *Integument*

In the decapod embryos, the surface of the embryo proper forms increasingly dense sets of very active cilia (Arnold and Williams-Arnold, 1980; Boletzky,

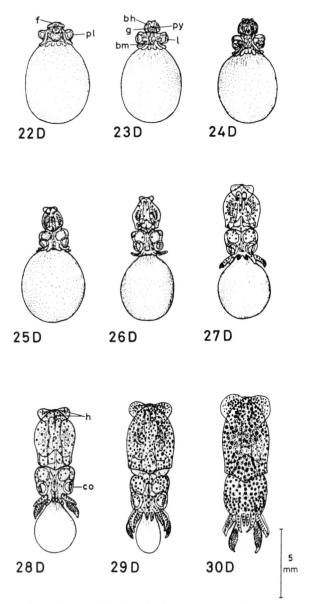

FIG. 4. Some embryonic stages of *Sepioteuthis lessoniana*, with embryos seen from the dorsal side (D), the staging 22 to 30 is according to Arnold (1965) (this corresponds roughly to Naef's stages XII to XX). f = fin, pl = primordium of lens, bh = branchial heart, g = gill, bm = buccal mass, py = posterior lobes of internal yolk sac, l = lens, h = Hoyle's organ (hatching gland), co = cornea (from Segawa, 1987). NB. Chromatophores appear first on ventral mantle surface at Stage 24 (c. stage XIV of Naef) according to Segawa (loc. cit.).

1980), whereas in octopod embryos, ciliation remains restricted to the outer yolk sac. Here a gradual reversal of the ciliary beat direction at early organogenetic stages leads to the peculiar first inversion described by Boletzky (1978–79), which has been observed in various species (Gabe, 1975; Joll, 1978; Nesis and Nigmatullin, 1978; Ambrose, 1981) but which does not occur in *Argonauta* embryos.

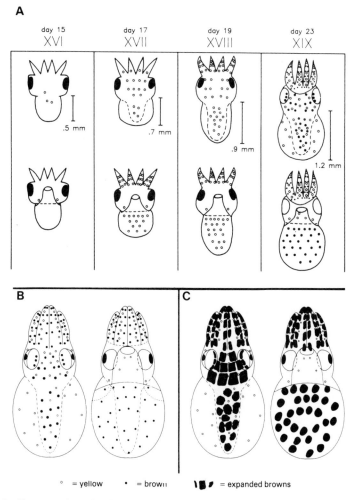

FIG. 5. Chromatophore development in *Octopus burryi*. (A) Embryos (without yolk sac) at stages XVI to XIX of Naef in dorsal view (upper row) and in ventral view (lower row). (B) Typical chromatophore arrangement at hatching (stage XX of Naef). (C) Typical dark body pattern of a live hatchling, with all brown chromatophores expanded and no yellows expanded (from Forsythe and Hanlon, 1985).

The early phase of chromatophore development in the integument belongs to the later part of organogenesis; chromatophores generally appear earlier in decapod embryos (Fig. 4) than in octopod embryos (Fig. 5) (see also Boletzky, 1984). The histological and ultrastructural aspects of chromatophore differentiation are now known in great detail thanks to the work on squid chromatophore development by Poggel and Fioroni (1986).

C. *Nervous System*

A particularly intriguing problem in cephalopod embryology is the developmental origin of the nervous system. Marquis (1981) has traced most ganglion rudiments in *Octopus vulgaris* from the stage of cellular immigration from the ectoderm; unfortunately this work has not yet been published. Marthy (1987) briefly reviews the results and includes some of the figures (Fig. 6).

D. *Gonads*

The problem of gonad differentiation during early stages of organogenesis is reviewed by Fioroni and Sundermann (1983). These authors confirm the

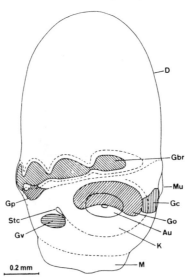

FIG. 6. Rudiments of the nervous system in an embryo of *Octopus vulgaris* at stage IX of Naef, represented in lateral view. Hatched areas represent the following ganglia: Gbr = brachial, Gp = pedal, Gv = visceral, Go = optic and Gc = cerebral ganglion. D = yolk, Mu = mouth, M = mantle, Au = eye, K = cephalic complex, Stc = statocyst (from Marquis, 1981).

results of earlier studies according to which the gonad rudiment is unpaired in decapod embryos, whereas in octopod embryos paired rudiments appear. The histological differentiation in relation to embryonic stages is very variable among different species or groups. (For sex differentiation at late embryonic and post-embryonic stages, see Richard and Lemaire, 1975; Lemaire and Richard, 1979.)

E. Digestive Tract

The basic process of early midgut differentiation from the mesendodermic ring (Marthy, 1982, 1985) is not yet entirely elucidated. As soon as the midgut rudiment is histologically distinct from the surrounding mesoderm, the further development leading to differentiation into the different parts of the digestive tract can easily be followed (Boletzky, 1978a). Lemaire et al. (1976) described the structural aspects of early digestive gland development in Sepia officinalis.

F. Funnel Complex

After closure of the ectodermic invaginations mentioned earlier, a pair of originally inconspicuous tissue streaks, which have split from the posterior ("ventral") arm rudiments (Boletzky, 1987a, 1988a,b), become prominent folds. These are the rudiments of the funnel tube, which are secondarily connected with another pair of folds produced by the pallio-visceral complex. The latter forms the funnel pouch or collar. Arnold et al. (1978) described the fine structure of the edges of the funnel tube rudiments that become fused to form a closed tube, the apex of which is finally re-opened by elimination of the closing membrane. In the Nautilus embryo (Fig. 7) this fusion never occurs, as the so-called hyponome is forming a functional funnel "tube" by means of overlapping margins (Arnold and Carlson, 1986; Arnold, 1987).

G. Arms and Tentacles

The individual arm rudiments become distinct within the arm crown anlage by the time of separation from the rudiments of the funnel tube. The exact number of arm ("tentacle") rudiments in Nautilus embryos at early organo-genetic stages remains to be established (Arnold and Carlson, 1986; Arnold, 1988). Within the coleoids, the decapod embryos of medium to large size

FIG. 7. Embryo of *Nautilus* seen from the left side, with only the uppermost part of the very large yolk mass shown (broken line). Note the curved shell indicating the beginning of exogastric coiling, one eye and tentacles below the mantle rim, and the hyponome at the ventral (posterior) side (right side of picture). (Courtesy of Prof. Dr. John M. Arnold, University of Hawaii.)

show five pairs of rudiments from the beginning of arm differentiation, the fourth pair (when counted from the buccal side backwards) becoming the specialized tentacles. In small embryos, such as those of ommastrephid squids (O'Dor *et al.*, 1982), some arm pairs are differentiated only in post-hatching stages (in addition to this, the ommastrephid tentacle rudiments fuse together as "rhynchoteuthion", at late embryonic stages, before the ventral arms are formed). In octopod embryos, there are always four pairs of arm rudiments, no matter what the size. The second decapodan pair (from the buccal side) of arm rudiments, which is modified in *Vampyroteuthis*, is most likely to be lacking in octopods (Boletzky, 1978–79). Sucker rudiments (see Nolte and Fioroni, 1983) always appear in a single series on each arm; biserial arrangement may then be achieved rather quickly by alternating enlargement and crowding, which may finally lead to a quadriserial arrangement in decapod embryos. In the embryos of certain incirrate and all cirrate octopods, the sucker rudiments remain uniserial (Boletzky, 1982a) (Fig. 1).

H. *Mantle*

In parallel with the funnel and arm differentiation, the mantle fold grows over the visceral complex to overlap finally with the funnel pouch. The mantle cavity thus formed encloses the gill rudiments and the so-called anal papilla (the anal opening being formed much later). Along with this growth process, the mantle surface overlying the typical coleoid shell sac forms a pair of fins from ear-like folds (Fig. 4). Their differentiation proceeds in close relation with the differentiation of the underlying shell complex (Bandel and Boletzky, 1979; Boletzky, 1982b). The early fin rudiments disappear in incirrate octopods (Boletzky, 1978–79); in contrast they grow very large in the embryo of *Heteroteuthis* (Boletzky, 1978b) and of cirrate octopods (Boletzky, 1982a).

The ectocochleate conditions of mantle development in *Nautilus* are described by Arnold (1988).

I. *Primary Lid*

The last great folding process in coleoid embryos is the formation of the primary lid. This process starts with the lengthening of the outer edges of arm bases III and IV in decapod embryos, and of arm bases II and III in octopod embryos. As mentioned above, this numbering does not take account of the homology of arm rudiments, because arm III of octopods is morphologically homologous to the tentacle (formed from arm bud IV) of decapods. The two pairs of posterior arm base projections form one anterior (dorsal) and one posterior (ventral) pair of folds reaching around the eyes. On each side, the complementary folds unite behind the eye and begin to form a membrane that progressively covers the eye complex (Arnold, 1984b). In all Sepioidea except *Spirula*, and in the myopsid Teuthoidea (loliginid and pickfordia-teuthid squids), it forms a closed cornea at late embryonic stages. A somewhat different type of cornea is formed in the octopods (overlapping edges of primary lid). In oegopsid squids and in *Spirula* the primary lid remains open in front of the eye lens.

J. *Eye*

The organogenesis of the eye complex excluding the primary lid comprises the eye vesicle (with the developing lens), the large optic ganglion and a cup-like ectodermic support. Its development has been described in *S. officinalis* by Lemaire and Richard (1978). Tissue culture experiments using this complex in *L. vulgaris* were carried out by Marthy and Aroles (1987). The ultrastructural differentiation of the retina cells in *Sepiella japonica* were analysed by Yamamoto (1985), and electrical responses of these developing cells were studied by Yamamoto *et al.* (1985).

K. *Body Axes*

For early embryonic stages, the so-called morphological or embryological orientation (emphasizing the basic molluscan bauplan) is often used. Organogenesis leads to the typical cephalopod organization, and for later embryonic stages, it thus becomes more natural to use the prospective "physiological" orientation for the description of organ positions (see Fioroni, 1978). Whereas in the morphological orientation, the mantle is above (dorsal) and the arm crown below (ventral), in the physiological orientation the arms are anterior or rostral the funnel tube is then ventral, and the mantle tip posterior or caudal. It should be kept in mind, however, that certain cephalopods, such as *Spirula*, do not normally swim in a position corresponding to the physiological orientation as defined above.

VII. Later Embryonic Stages

There is no distinct "end" to organogenesis proper, so that arbitrary criteria must be used where post-gastrular development is to be divided into an organogenetic and a post-organogenetic phase. Depending on the particular focus of a study, these criteria will be concerned either with the overall aspect of the embryo, or with the state of differentiation of a given organ or organ complex. In the former case, a basic difference lies in whether living or preserved embryos are under consideration.

A. *Yolk Sac*

In the living embryo the size and shape of the inner yolk sac provide a helpful guide in the assessment of organ development, because they show the "mould" of tissue concentration due to different organ complexes. The inner yolk sac is actually compressed and its contents are partly extruded into the outer yolk sac by the steadily contracting organs of the embryo. Exceptions to this are the embryos of ommastrephid squids where the rudimentary outer yolk sac is progressively reduced in size during the early stages of organogenesis. The yolk mass thus forms a voluminous core whose volume is reduced only at late embryonic stages (O'Dor *et al.*, 1982). In most of the other cephalopod embryos, it is generally around stage XV of Naef (1928) that the inner yolk sac reaches its minimal size. At this stage all organ complexes are lying in their definitive positions relative to each other; only the primary lid and the dorsal "closure" of the arm crown are still incomplete (Boletzky, 1978–79). As the subsequent stages are characterized by linear body growth accompanied by further histological differentiations, stage XV provides a good landmark to separate organogenesis proper from organ growth and differentiation leading to fully functional states. In nearly all known cephalopods, this late phase starts with an increase of the inner yolk sac size. With the exception of the octopus *Eledone moschata* (see Boletzky, 1978–79), the yolk is shifted from the outer to the inner sac; the latter will provide a nutrient reserve immediately after hatching, without interfering with the definitive, functional digestive organs.

B. *Integument*

Certain processes of differentiation are largely independent of the onset of embryonic growth. Integumental structures show particularly a wide variety of heterochronic shifts relative to other organogenetic events (Fioroni, 1982a). The structural processes of chromatophore differentiation in the skin (Poggel and Fioroni, 1986) demonstrate the complex nature of the ontogenesis of colour patterning, which attains its climax only at advanced juvenile stages in most cephalopods (one of the exceptions being *Sepia* where hatchling patterns are already very elaborate). At late embryonic stages, the generally simple chromatophore patterns appear to be species-specific (see Segawa *et al.*, 1988). Marthy and Hanlon (1983) describe a method of selective chromatophore destruction by laser micro-irradiation that may be useful in studying chromatophore pattern formation and repair in the skin of embryonic and juvenile cephalopods.

The integumental surface undergoes important modifications during later embryonic stages. In decapod embryos, the simple pattern of ciliary tufts undergoes rearrangements on the mantle surface, with differentiation of long ciliary bands (Arnold and Williams-Arnold, 1980; Boletzky, 1980). The cilia of the cells forming these bands are shorter than those of the tuft cells, and they can in fact be recognized before they are arranged in bands (Boletzky, 1982c). The long cilia of the tuft cells are also different due to their tendency to form a characteristic fixation artefact known as "paddle cilia" (Boletzky, 1980). Sundermann-Meister (1978) has described a type of ciliated cells in the skin of late embryonic and early post-embryonic squid (*L. vulgaris*) which is likely to have a receptor function. The ultrastructure of this type of ciliary cell in both *L. vulgaris* and *S. officinalis* is described by the same author (Sundermann, 1982, 1983) (Fig. 8).

In incirrate octopod embryos, the skin development in most species is characterized by the differentiation of special organs, named organs of Kölliker (Brocco *et al.*, 1974; Boletzky, 1978–79; Fioroni, 1982a). Their role in hatching will be mentioned below. Fioroni (1982a) provides a comparative

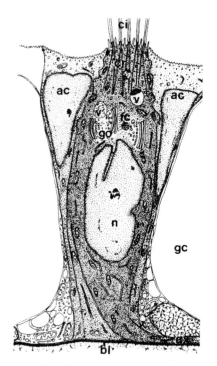

FIG. 8. Schematic representation of a section across an epidermal line of ciliated cells in hatchling *Loligo vulgaris*; ac = accessory cell, bl = basal lamina, ci = cilia, cr = ciliary rootlets, fc = fibrous cluster, gc = goblet cell, go = Golgi complex, n = nucleus, v = large vesicle (from Sundermann, 1983).

description of skin (including the hatching gland) and sucker development in three different incirrate octopods, two representatives of the benthic octopodids (*Octopus vulgaris, Eledone cirrosa*) and one member of the pelagic argonautids (*Argonauta argo*), with a discussion on the relation between these developmental features and the post-embryonic mode of life. In the embryos of cirrate octopods the skin does not show any trace of Kölliker organs at advanced stages (Boletzky, 1982a). The suckers in cirrate embryos develop from a single file of rudiments on each arm, the cirri appearing only during post-embryonic development (Boletzky, 1978–79).

C. Musculature

The ultrastructure of developing muscle cells in the cuttlefish *S. japonica* is described by Matsuno (1987) who defines two late embryonic and one early post-embryonic phase of muscle cell differentiation in the circular mantle muscle.

D. *Statocysts*

Culture studies carried out in closed seawater systems have brought to light a behavioural defect in hatchlings of several cephalopod species that is clearly related to malformations of some or all of the statocyst elements, apparently as the result of environmental causes (Colmers *et al.*, 1984).

E. *Perivitelline Fluid*

Although cephalopod embryos can develop normally in sterile seawater (Marthy, 1978–79), natural conditions are characterized by a chemical composition of the perivitelline fluid different from seawater. In the squid *L. vulgaris* the perivitelline fluid at late embryonic stages has been shown to contain a tranquillizing factor which clearly minimizes the incidence of premature hatching (Marthy *et al.*, 1976). The chemical and physiological properties of this natural tranquillizer have been studied by Weischer and Marthy (1983) who suggested that the effective part is a polypeptide or a protein of about 60 kilodalton.

F. *Hatching*

As mentioned earlier, the period from laying to hatching varies with temperature (Fig. 9). The mechanism of hatching in squids has been shown to depend on the coordinated action of the hatching gland enzyme (Denucé and Formisano, 1982) which dissolves locally the chorion and surrounding envelopes, and of the integumental ciliary apparatus by means of which the animal moves through the "tunnel" opened by the enzyme action (Boletzky, 1979). In *Sepia* embryos the ciliary bands, typical of the hatchling squid mantle, are differentiated only in close relation with the hatching gland, while in the sepiolids the integument only carries the ciliary tufts. The mantle apex of sepiolid hatchlings is equipped with a tough spine-like organ which is associated with the hatching gland (Boletzky, 1982c).

In the incirrate octopods the embryonic integument is devoid of motile cilia. The Kölliker organs provide a guiding ("one way") structure during hatching (Boletzky, 1978–79).

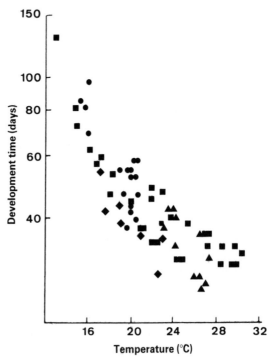

Fɪɢ. 9. The relationship between temperature and development time in octopuses with planktonic young. ● = *Octopus bimaculatus*, ■ = *Octopus vulgaris*, ▲ = *Octopus cyanea*, ◆ = *Octopus tetricus* (from Ambrose, 1981, based on different sources, including Joll, 1978).

VIII. Post-hatching Development

Our current knowledge on post-hatching development in cephalopods has been summarized very recently (Okutani, 1987; Boletzky, 1987d; Vecchione, 1987). A workshop organized in 1985 (under the auspices of the Cephalopod International Advisory Council) dealt entirely with "Early Growth Stages of Cephalopods" and resulted in a handbook published by the Smithsonian Institution in Washington, D.C. (now in press). In combination with this workshop, a symposium on "Biology and Distribution of Early Juvenile Cephalopods" was held (see Mangold and Boletzky, 1985). Thus it is quite clear that post-hatching development of cephalopods is receiving much attention, in particular with regard to its importance in biological oceanography, fishery biology and small-scale culture. Cephalopods are still potential material for industrial aquaculture, but as long as the seafood market is provisioned with cephalopods fished from the sea, there is no urgent need for large-scale ventures. Thus the main demand for cultured cephalopods is in science, especially for biomedical research in the widest sense of the term (see Boletzky and Hanlon, 1983). The role of cephalopods as "research models" is widely recognized, but generally it is the adult animal that has to provide whole organs (e.g. the eye) or isolated cells (e.g. the giant axons). In culture work aiming only at the provision of such "source" animals, the early developmental stages would only be of interest because they pose the conditions for later success or failure. Ignoring the requirements of developing embryos and newly hatched animals with regard to water quality and temperature conditions will inevitably lead to failure in culture work.

Successful culture work, on the other hand, is based on a keen interest in the animal throughout its life cycle, including the embryonic phase. During the post-hatching to adult development stage, behavioural aspects are of particular importance, especially feeding (cf. Mauris, 1988). These aspects are of equal importance to ecological studies which provide insights that ultimately serve the interests of fishery biology (cf. Voss, 1983). Whatever the main interest is, behavioural studies on newly hatched cephalopods should take account of the immediately preceding life stage, i.e. late embryonic development and the process of hatching itself.

Here is one of the most serious gaps in our knowledge of cephalopod development. It is true that the "hatching stage" is ill-defined due to the continuous perfection of the young animal in late embryonic life and the generalized persistence of yolk reserves (Boletzky, 1987a), but this should not be taken to mean that no thorough physiological change is imposed on the organism at the moment of hatching. The "perinatal" biology of cephalopods should be given special attention in future research work.

IX. Conclusions and Perspective

In limiting the literature survey of this review to publications that have appeared since 1975 (i.e. those papers published since the bibliography of Fioroni's 1978 book was closed), we deliberately focus our attention on the most recent research activities concerned with the embryonic phase of cephalopod life. Thus no mention is made of a host of great studies that form the indispensable basis of our actual research—this is the only way to realize where current interest appears to be concentrated.

At first sight, the chapter headings and subheadings suggest a rather coherent coverage of all stages of embryonic development based on the study of most of the organ systems, but a closer look will modify this impression. Clearly some stages and some organ systems are more intensively studied than others, which is not astonishing considering the small number of scientists working in this field.

A major gap becomes strikingly clear considering the wealth of studies dealing with the adult nervous system of cephalopods. Apart from the tissue culture experiments on the oculo-gangliar complex of squid embryos (Marthy, 1987; Marthy and Aroles, 1987), there is only one study dealing with the embryonic development of the nervous system of a cephalopod (Marquis, 1981). Preparations for its publication in vol. 99 of the journal *Verhandlungen der Naturforschenden Gesellschaft, Basel* are underway (Marquis, personal communication, 1988). On the other hand, there is a considerable amount of recent work concerned with the surface morphology of cephalopod embryos, which may be seen in relation to the increasingly generalized utilization of scanning electron microscopes in descriptive embryology (as in juvenile morphology).

As to experimental embryology, Naef (1928) has already stressed the suitability of cephalopod embryos for experimental work. The whole body of information accumulated in this field during the past half century since Naef published his monograph was reviewed by Marthy (1978–79). The same author is preparing a comprehensive overview of experimental cephalopod embryology (Marthy, personal communication, 1988). This field of research is clearly underrated by many developmental biologists who could profit by the topological simplicity of the blastulation pattern in cephalopods, which contrasts with the spiralian mode of other molluscs. It is interesting to note that one of the promoters of modern experimental embryology of cephalopods is now involved in descriptive embryology set against an evolutionary background; his continuing work on the morphology of *Nautilus* embryos (Arnold and Carlson, 1986; Arnold, 1988) is providing direct evidence of homologous patterns in the embryogenesis of very distant forms. For example, the striking similarity between the early shell (protoconch) structure in *Nautilus* and *Sepia*, especially with regard to the so-called cicatrix, which was noted by Bandel and Boletzky (1979), is now confirmed by the observations of Arnold (1988; see also Arnold et al., 1987).

Questions raised by comparative/evolutionary embryology, following the tradition of descriptive developmental morphology, are truly stimulating to the field of experimental embryology, and *vice versa*. However, experimental studies are generally possible with only a limited number of "models", which, in the case of cephalopods, appear to be embryos of medium to small size. On the other hand, some of the most intriguing questions in cephalopod biology are related to those forms that produce eggs of extremely large size. In terms of volume (c. 2000 mm^3), these eggs are four orders of magnitude above the smallest cephalopod eggs (0.2 mm^3). In terms of developmental time, the variation ranges from one year in *Bathypolypus arcticus* (O'Dor and Macalaster, 1983) [and possibly more than one year in cirrate octopods (cf. Boletzky, 1982a)], to about one week (Boletzky, 1987a). Needless to say, the latter category, which includes developmental times up to several weeks depending on environmental temperature, is the most attractive for experimental work, apart from the fact that very large eggs are rarely found in numbers sufficient for standardized serial experiments.

Knowledge about the "rare" forms, which generally lie outside the scope of experimental embryology, is due in general to the work achieved during oceanographic expeditions and related activities of biologists working in marine stations, including those running public aquaria (cf. Arnold and Carlson, 1986). A hope to fill the most tantalizing gaps in our knowledge of comparative cephalopod embryology—be it for purely embryological aspects or for a better understanding of all aspects of life-history adaptation— resides in the continuation of efforts made on a broad basis of biological oceanography and marine biology. It is to be hoped that 10 years from now the embryonic development of *Spirula* and of *Vampyroteuthis*, especially with regard to cleavage patterns and shell sac differentiation, will be known. This goal can be reached if the eggs are recognized in plankton samples, or adult animals are captured in sufficiently good condition to allow at least the collection of mature ovarian eggs and spermatozoids for artificial fertilization. Ultimately even normal spawning under large-scale aquarium conditions (cf. O'Dor *et al.*, 1982) should be possible with these delicate deep-sea animals.

Acknowledgements

The many helpful suggestions of Prof. P. Fioroni are gratefully acknowledged. I also thank authors for permission to reproduce figures from their published or unpublished work.

References

Ambrose, R. F. (1981). Observations on the embryonic development and early post-embryonic behavior of *Octopus bimaculatus*. *Veliger* **24**, 139–146.

Arnold, J. M. (1965). Normal embryonic stages of the squid, *Loligo pealii* (Lesueur). *Biological Bulletin* **128**, 24–32.

Arnold, J. M. (1976). Cytokinesis in animal cells: new answers to old questions. *In* "The Cell Surface in Animal Embryogenesis and Development" (G. Poste and G. L. Nicholson, eds), pp. 55–80. Elsevier/North-Holland Biomedical Press, Amsterdam.

Arnold, J. M. (1984a). Cephalopods. *In* "The Mollusca", Vol. 7, Reproduction, pp. 419–454, Academic Press, London.

Arnold, J. M. (1984b). Closure of the squid cornea: A muscular basis for embryonic tissue movement. *Journal of Experimental Zoology* **232**, 187–195.

Arnold, J. M. (1987). Reproduction and embryology of *Nautilus*. *In* "Nautilus" (W. B. Saunders and N. H. Landman, eds.), pp. 353–372. Plenum Publishing Corporation, New York.

Arnold, J. M. (1988). Some observations on the cicatrix of *Nautilus* embryos. *In* "Cephalopods: Present and Past" (J. Wiedmann and J. Kullmann, eds). pp. 181–190. Schweizerbart'sche Verlagsbuchhandlung, Stuttgart.

Arnold, J. M. and Carlson B. A. (1986). Living *Nautilus* embryos: preliminary observations. *Science* **232**, 73–76.

Arnold, J. M. and Williams-Arnold, L. D. (1976). The egg cortex problem as seen through the squid eye. *American Zoologist* **16**, 421–446.

Arnold, J. M. and Williams-Arnold, L. D. (1980). Development of the ciliature pattern on the embryo of the squid *Loligo pealei*: a scanning electron microscope study. *Biological Bulletin* **159**, 102–116.

Arnold, J. M., Landman, N. H. and Mutvei, H. (1987). Development of the embryonic shell of *Nautilus*. *In* "Nautilus" (W. B. Saunders and N. H. Landman, eds.), pp. 373–400. Plenum Publishing Corporation, New York.

Arnold, J. M., Williams-Arnold, L. D. and Peters, V. (1978). Fusion of tissue masses in embryogenesis. *Developmental Biology* **65**, 155–170.

Bandel, K. and Boletzky, S. v. (1979). A comparative study of the structure, development and morphological relationships of chambered cephalopod shells. *Veliger* **21**, 313–354.

Bergstrom, B. and Summers, W. C. (1983). *Sepietta oweniana*. *In* "Cephalopod Life Cycles" (P. R. Boyle ed.), Vol. I, pp. 75–91. Academic Press, London.

Boletzky, S. v. (1978a). Gut development in cephalopods: a correction. *Revue suisse de Zoologie* **85**, 379–380.

Boletzky, S. v. (1978b). Premières données sur le développement embryonnaire du Sepiolidé pélagique *Heteroteuthis* (Mollusca, Cephalopoda). *Haliotis* **9**, 81–84.

Boletzky, S. v. (1978–79). Nos connaissances actuelles sur le développement des octopodes. *Vie et Milieu* **28–29 AB**, 85–120.

Boletzky, S. v. (1979). Ciliary locomotion in squid hatching. *Experientia* **35**, 1051–1052.

Boletzky, S. v. (1980). Les "cils à raquette" chez les embryons des Céphalopodes—Cas de *Loligo vulgaris*. *Bulletin de la Société Zoologique de France* **105**, 209–214.

Boletzky, S. v. (1982a). On eggs and embryos of cirromorph octopods. *Malacologia* **22**, 197–204.

Boletzky, S. v. (1982b). Developmental aspects of the mantle complex in coleoid cephalopods. *Malacologia* **23**, 165–175.

Boletzky, S. v. (1982c). Structure tégumentaire de l'embryon et mode d'éclosion chez les Céphalopodes. *Bulletin de la Société Zoologique de France* **107**, 475–482.

Boletzky, S. v. (1983a). *Sepia officinalis*. *In* "Cephalopod Life Cycles" (P. R. Boyle ed.), Vol. I. pp. 31–52. Academic Press, London.

Boletzky, S. v. (1983b). *Sepiola robusta*. *In* "Cephalopod Life Cycles" (P. R. Boyle ed.), Vol. I, pp. 53–67. Academic Press, London.

Boletzky, S. v. (1984). The embryonic development of the octopus *Scaeurgus unicirrhus* (Mollusca: Cephalopoda)—Additional data and discussion. *Vie et Milieu* **34**, 87–93.

Boletzky, S. v. (1986). Encapsulation of cephalopod embryos: A search for functional correlations. *American Malacological Bulletin* **4**, 217–227.

Boletzky, S. v. (1987a). Embryonic phase. *In* "Cephalopod Life Cycles" (P. R. Boyle ed.), Vol. II, pp. 5–31. Academic Press, London.

Boletzky, S. v. (1987b). Ontogenetic and phylogenetic aspects of the cephalopod circulatory system. *Experientia* **43**, 478–483.

Boletzky, S. v. (1987c). On egg and capsule dimensions in *Loligo forbesi* (Mollusca: Cephalopoda): a note. *Vie et Milieu* **37**, 187–192.

Boletzky, S. v. (1987d). Juvenile behaviour. *In* "Cephalopod Life Cycles" (P. R. Boyle ed.), Vol. II, pp. 45–60. Academic Press, London.

Boletzky, S. v. (1988a). Cephalopod development and evolutionary concepts. *In* "The Mollusca", Vol. 12: Paleontology and Neontology of Cephalopods, pp. 185–202, Academic Press, London.

Boletzky, S. v. (1988b). Characteristics of cephalopod embryogenesis. *In* "Cephalopods—Present and Past" (J. Wiedmann and J. Kullmann, eds.), pp. 167–179. Schweizerbart'sche Verlagsbuchhandlung, Stuttgart.

Boletzky, S. v. and Centelles, J. (1978–79). *Argonauta argo* (Mollusca: Cephalopoda) dans la région de Banyuls-sur-Mer. *Vie et Milieu*, **28–29 AB**, 659–660.

Boletzky, S. v. and Hanlon, R. T. (1983). A review of the laboratory maintenance, rearing and culture of cephalopod molluscs. *Memoirs of the National Museum Victoria* **44**, 147–187.

Boyle, P. R. (Ed.) (1983). "Cephalopod Life Cycles", Vol. I, Species Accounts, 475 pp. Academic Press, London.

Boyle, P. R. (Ed.) (1987). "Cephalopod Life Cycles", Vol. II, Comparative Reviews, 441 pp. Academic Press, London.

Brocco, S. L., O'Clair, R. M. and Cloney, R. A. (1974). Cephalopod integument: the ultrastructure of Kölliker's organs and their relationship to setae. *Cell and Tissue Research* **151**, 293–308.

Cartwright, J. and Arnold, J. M. (1980). Intercellular bridges in the embryo of the Atlantic squid, *Loligo pealei* I. Cytoplasmic continuity and tissue differentiation. *Journal of Embryology and Experimental Morphology* **57**, 3–24.

Cartwright, J. and Arnold, J. M. (1981). Intercellular bridges in the embryo of the Atlantic squid, *Loligo pealei*. II: Formation of the Bridge. *Cell Motility* **1**, 455–468.

Clarke, M. R. (1988). Evolution of recent cephalopods—A brief review. *In* "The Mollusca", Vol. 12: Paleontology and Neontology of Cephalopods, pp. 331–340. Academic Press, London.

Colmers, W. F., Hixon, R. F., Hanlon, R. T., Forsythe, J. W., Ackerson, M. V., Wiederhold, M. L. and Hulet, W. H. (1984). "Spinner" cephalopods: defects of statocyst suprastructures in an invertebrate analogue of the vestibular apparatus. *Cell and Tissue Research* **236**, 505–515.

Crawford, K. (1985). Pronuclear migration of *in vitro* fertilized and activated eggs of the squid *Loligo pealei* (abstract), *Biological Bulletin* **169**, 540.

Denucé, J. M. and Formisano, A. (1982). Circumstantial evidence for an active contribution of Hoyle's gland to enzymatic hatching of Cephalopod embryos. *Archives Internationales de Physiologie et de Biochimie* **90 B**, 185–186.

Fioroni, P. (1977). Die Entwicklungstypen der Tintenfische. *Zoologische Jahrbücher*, Abt. *Anatomie* **98**, 441–475.

Fioroni, P. (1978). Cephalopoda. *In* "Morphogenese der Tiere" (F. Seidel ed.), Lieferung 2: G5-I, 181 pp. Gustav Fischer Verlag, Jena.

Fioroni, P. (1979a). Phylogenetische Abänderungen der Gastrula bei Mollusken. *In* "Ontogenese und Phylogenie" (R. Siewing ed.), pp. 82–100. Paul Parey Verlag, Hamburg und Berlin.

Fioroni, P. (1979b). Abänderungen des Gastrulationsverlaufs und ihre phylogenetische Bedeutung. *In* "Ontogenese und Phylogenie" (R. Siewing ed.), pp. 101–119. Paul Parey Verlag, Hamburg und Berlin.

Fioroni, P. (1980). Ontogenie—Phylogenie. Eine Stellungnahme zu einigen neuen entwicklungsgeschichtlichen Theorien. *Zeitschrift für zoologische Systematik und Evolutionsforschung* **18**, 90–103.

Fioroni, P. (1982a). Zur Epidermis- und Saugnapfentwicklung bei Octopoden, ein entwicklungsgeschichtlicher Vergleich. *Revue suisse de Zoologie* **89**, 355–374.

Fioroni, P. (1982b). Larval organs, larvae, metamorphosis and types of development of Mollusca—a comprehensive review. *Zoologische Jahrbücher* Abt. *Anatomie*, **108**, 375–420.

Fioroni, P. (1982c). Entwicklungstypen und Schlüpfstadien bei Mollusken—einige allgemeine Befunde. *Malacologia* **22**, 601–609.

Fioroni, P. (1982d). Allgemeine Aspekte der Mollusken–Entwicklung. *Zoologische Jahrbücher*, Abt. *Anatomie* **107**, 85–121.

Fioroni, P. (1983). Phylogenetische Aspekte der Furchungsmuster. *Revue suisse de Zoologie* **90**, 939–949.

Fioroni, P. and Sundermann, G. (1983). Zur embryonalen Differenzierung der Gonadenanlage bei coleoiden Cephalopoden. *Zoologische Jahrbücher*, Abt. *Anatomie* **109**, 153–165.

Forsythe, J. W. and Hanlon, R. T. (1985). Aspects of egg development, post-hatching behavior, growth and reproductive biology of *Octopus burryi* Voss, 1950 (Mollusca: Cephalopoda). *Vie et Milieu* **35**, 273–282.

Gabe, S. H. (1975). Reproduction in the giant octopus of the North Pacific, *Octopus dofleini martini*. *Veliger* **18**, 146–150.

Gomi, F., Yamamoto, M. and Nakazawa, T. (1986). Swelling of egg during development of the cuttlefish, *Sepiella japonica*. *Zoological Science* **3**, 641–645.

Hixon, R. F. (1983). *Loligo opalescens*. *In* "Cephalopod Life Cycles" (P. R. Boyle ed.), Vol. I. pp. 95–114. Academic Press, London.

Joll, L. M. (1978). Observations on the embryonic development of *Octopus tetricus* (Mollusca: Cephalopoda). *Australian Journal of Marine and Freshwater Research* **29**, 19–30.

Kao, K. R. (1985). Blastodisc formation in squid (*Loligo pealei*) eggs: a possible role for micro-tubules (abstract). *Biological Bulletin* **169**, 541.

Lemaire, J. (1970). Table de développement embryonnaire de *Sepia officinalis* L. (Mollusque Céphalopode). *Bulletin de la Société Zoologique de France* **95**, 773–782.

Lemaire, J. and Richard, A. (1978). Organogenèse de l'oeil du Céphalopode *Sepia officinalis* L. *Bulletin de la Société Zoologique de France* **103**, 373–377.

Lemaire, J. and Richard, A. (1979). Présomption d'un contrôle endocrine de la différenciation du sexe chez la seiche embryonnaire. *Annales d'Endocrinologie* (Paris) **40**, 91–92.

Lemaire, J., Richard, A. and Decleir, W. (1976). Le foie embryonnaire de *Sepia officinalis* L. (Mollusque Céphalopode). I–Organogenèse. *Haliotis* **6**, 287–296.

Mangold, K. (1987). Reproduction. *In* "Cephalopod Life Cycles" (P. R. Boyle ed.), Vol. II, pp. 157–200. Academic Press, London.

Mangold, K. and Boletzky, S. v. (1985). Preface. *In* "Biology and Distribution of Early Juvenile Cephalopods" (K. Mangold and S. v. Boletzky, eds). *Vie et Milieu* **35**, 137–138.

Marquis, F. (1981). Untersuchungen über die Entwicklung des Nervensystems im

Embryo von *Octopus vulgaris* Lam. Doctoral dissertation, University of Basel, 76 pp.

Marthy, H. J. (1976). Les déterminismes dans la morphogenèse—Contribution à l'embryologie expérimentale des Céphalopodes. Thèse de Doctorat d'Etat, University Paris VI, No. enregistr. C.N.R.S. AO 12426, 55 pp.

Marthy, H. J. (1978–79). Embryologie expérimentale chez les Céphalopodes. *Vie et Milieu* **28–29 AB**, 121–142.

Marthy, H. J. (1982). The cephalopod egg, a suitable material for cell and tissue interaction studies. *In* "Embryonic Development", Part B: Cellular Aspects, pp. 223–233. Alan R. Liss, Inc., New York.

Marthy, H. J. (1985). Morphological bases for cell-to-cell and cell-to-substrate interaction studies in cephalopod embryos. *In* "Cellular and Molecular Control of Direct Cell Interactions" (H. J. Marthy ed.), pp. 159–197. Plenum Publishing Corporation, New York.

Marthy, H. J. (1987). Ontogenesis of the nervous system in cephalopoids. *In* "Nervous Systems in Ivertebrates" (M. A. Ali, ed.), pp. 443–459, Plenum Publishing Corporation, New York.

Marthy, H. J. and Aroles, L. (1987). *In vitro* culture of embryonic organ and tissue fragments of the squid *Loligo vulgaris* with special reference to the establishment of a long term culture of ganglion-derived nerve cells. *Zoologische Jahrbücher,* Abt. *Physiologie* **91**, 189–202.

Marthy, H. J. and Hanlon, R. T. (1983). Effects of laser microirradiation in chromatophore organs of embryonic and juvenile cephalopods. *Mikroskopie* **40**, 35–40.

Marthy, H. J., Hauser, R. and Scholl, A. (1976). Natural tranquilliser in cephalopod eggs. *Nature* **261**, 496–497.

Matsuno, A. (1987). Ultrastructural studies on developing oblique-striated muscle cells in the cuttlefish, *Sepiella japonica* Sasaki. *Zoological Science* **4**, 53–59.

Mauris, M. E. (1988). Le comportement prédateur de la Sépiole *Sepiola affinis*— Approches expérimentales en éco-éthologie. Doctoral dissertation, University of Paris VI, 108 pp.

Meister, G. (1977). Untersuchungen an vakuolisierten Rundzellen im Blut von Embryonen verschiedener Tintenfisch–Arten (Mollusca, Cephalopoda). *Zoologische Jahrbücher,* Abt. *Anatomie,* 97, 54–67.

Misaki, H. and Okutani, T. (1976). Studies on early life history of decapodan mollusca–VI. An evidence of spawning of an oceanic squid, *Thysanoteuthis rhombus* Troschel, in the Japanese Waters. *"Venus", the Japanese Journal of Malacology* **35**, 211–213.

Naef, A. (1928). Die Cephalopoden. *Fauna und Flora des Golfes von Neapel* **35 (2)**, pp. 1–357, Friedländer & Son, Berlin.

Natsukari, Y. (1970). Egg-laying behaviour, embryonic development and hatched larva of the Pygmy Cuttlefish, *Idiosepius pygmaeus paradoxus* Ortmann. *Bulletin of the Faculty of Fisheries, Nagasaki University* **30**, 15–29.

Natsukari, Y. (1976). SCUBA diving observations on the spawning ground of the squid, *Doryteuthis kensaki* (Wakyia and Ishikawa, 1921) (Cephalopoda: Loliginidae). *"Venus", the Japanese Journal of Malacology* **35**, 206–208.

Natsukari, Y. (1979). Discharged, unfertilized eggs of the cuttlefish, *Sepia (Metasepia) tullbergi. Bulletin of the Faculty of Fisheries, Nagasaki University* **47**, 27–28.

Nesis, K. N. and Nigmatullin, Ch. M. (1978). A record of egg-masses of the bottom octopus *Eledone caparti* (Octopodidae) in the stomachs of blue sharks. *Zoological Journal (Zool. Zh.)* **57**, 1324–1329.

Nicol, J. A. C. (1967). "The Biology of Marine Animals". Pitman & Sons Ltd, London, 699 pp.

Nolte, K. and Fioroni, P. (1983). Zur Entwicklung der Saugnäpfe bei coleoiden Tintenfischen. *Zoologischer Anzeiger* **211**, 329–340.

O'Dor, R. K. (1983). *Illex illecebrosus. In* "Cephalopod Life Cycles" (P. R. Boyle, ed.), Vol. I, pp. 175–199. Academic Press, London.

O'Dor, R. K. and Macalaster, E. G. (1983). *Bathypolypus arcticus. In* "Cephalopod Life Cycles" (P. R. Boyle, ed.), Vol. I, pp. 401–410. Academic Press.

O'Dor, R. K. and Webber, D. M. (1986). The constraints on cephalopods: why squid aren't fish. *Canadian Journal of Zoology* **64**, 1591–1605.

O'Dor, R. K., Balch, N., Foy, E. A., Hirtle, R. W. M., Johnston, D. A. and Amaratunga, T. (1982). Embryonic development of the squid, *Illex illecebrosus*, and effect of temperature on developmental rates. *Journal of North-west Atlantic Fishery Science* **3**, 41–45.

Okiyama, M. and Kasahara, S. (1975). Identification of the so-called "common squid eggs" collected in the Japan Sea and adjacent waters. *Bulletin of the Japan Sea Regional Fisheries Research Laboratory* **26**, 35–40.

Okutani, T. (1978). Studies on early life history of decapodan mollusca–VII. Eggs and newly hatched larvae of *Sepia latimanus* Quoy & Gaimard. *"Venus", the Japanese Journal of Malacology*, **37**, 245–248.

Okutani, T. (1983). *Todarodes pacificus. In* "Cephalopod Life Cycles" (P. R. Boyle, ed.), Vol. I, pp. 201–214. Academic Press, London.

Okutani, T. (1987). Juvenile morphology. *In* "Cephalopod Life Cycles" (P. R. Boyle, ed.), Vol. II, pp. 33–44. Academic Press, London.

Pickford, G. E. (1949). The distribution of the eggs of *Vampyroteuthis infernalis* Chun. *Sears Foundation Journal of Marine Research* **8**, 73–83.

Poggel, M. and Fioroni, P. (1986). Die histologische und ultrastrukturelle Differenzierung der Chromatophoren von *Loligo vulgaris* Lam. (Cephalopoda). *Zoologischer Anzeiger* **217**, 207–227.

Richard, A. and Lemaire, J. (1975). Détermination et différenciation sexuelles chez la seiche *Sepia officinalis* L. (Mollusque Céphalopode). *Pubblicazioni della Stazione Zoologica di Napoli* 39 suppl., 574–594.

Roper, C. F. E. and Sweeney, M. J. (1983). Techniques for fixation, preservation, and curation of cephalopods. *Memoirs of the National Museum Victoria* **44**, 29–47.

Sabirov, R. M., Arkhipkin, A. I., Tsygankov, V. Yu. and Shchetinnikov, A. S. (1987). Egg-laying and embryonal development of Diamond-shaped Squid *Thysanoteuthis rhombus* (Oegopsida, Thysanoteuthidae). *Zoological Journal (Zool. Zh.)* **66**, 1155–1163.

Segawa, S. (1987). Life history of the Oval Squid, *Sepioteuthis lessoniana* in Kominato and adjacent waters, central Honshu, Japan. *Journal of the Tokyo University of Fisheries* **74**, 67–105.

Segawa, S., Yang, W. T., Marthy, H.-J. and Hanlon, R. T. (1988). Illustrated embryonic stages of the Eastern Atlantic squid *Loligo forbesi*. *Veliger* **30**, 230–243.

Segmüller, M. and Marthy, H.-J. (1984). The cephalopod embryo, a suitable system for the *in situ* recording of cell migration. *Experientia* 636.

Singley, C. T. (1977). An Analysis of Gastrulation in *Loligo pealei*. Doctoral dissertation, University of Hawaii, 163 pp.

Singley, C. T. (1983). *Euprymna scolopes. In* "Cephalopod Life Cycles" (P. R. Boyle, ed.), Vol. I, pp. 69–74. Academic Press, London.

Summers, W. C. (1983). *Loligo pealei. In* "Cephalopod Life Cycles" (P. R. Boyle, ed.), Vol. I, pp. 115–142. Academic Press, London.

Sundermann, G. (1980). Die Ultrastruktur der vakuolisierten Rundzellen von *Loligo vulgaris* Lam. (Mollusca, Cephalopoda). *Zoologische Jahrbücher*, Abt. *Anatomie* 103, 93–104.

Sundermann, G. (1982). Untersuchungen an den Cilienzell-Linien auf Armen und Kopf bei Cephalopoden. *Mitteilungen der deutschen malakologischen Gesellschaft* 3 suppl., 61–63.

Sundermann, G. (1983). The fine structure of epidermal lines on arms and head of postembryonic *Sepia officinalis* and *Loligo vulgaris* (Mollusca, Cephalopoda). *Cell and Tissue Research* 232, 669–677.

Sundermann-Meister, G. (1978). Ein neuer Typ von Cilienzellen in der Haut von spätembryonalen und juvenilen *Loligo vulgaris* (Mollusca, Cephalopoda). *Zoologische Jahrbücher*, Abt. *Anatomie* 99, 493–499.

Suzuki, S., Misaki, H. and Okutani, T. (1979). Studies on early life history of decapodan mollusca–VIII. A supplementary note on floating egg masses of *Thysanoteuthis rhombus* Troschel in Japan—The first underwater photography. *"Venus", the Japanese Journal of Malacology* 38, 153–155.

Vecchione, M. (1987). Juvenile ecology. *In* "Cephalopod Life Cycles" (P. R. Boyle, ed.), Vol. II, pp. 61–84. Academic Press, London.

Voss, G. L. (1977). Classification of recent cephalopods. *Symposia of the Zoological Society London* 38, 575–579.

Voss, G. L. (1983). A review of cephalopod fisheries biology. *Memoirs of the National Museum Victoria* 44, 229–241.

Weischer, M.-L. and Marthy, H.-J. (1983). Chemical and physiological properties of the natural tranquilliser in the cephalopod eggs. *Marine Behaviour and Physiology* 9, 131–138.

Wittland, C. and Fioroni, P. (1982). Zum ontogenetischen Auftreten von ectodermalen Vesikeln bei dibranchiaten Cephalopoden. *Zoologische Beiträge*, Neue Folge 28, 67–77.

Worms, J. (1983). *Loligo vulgaris. In* "Cephalopod Life Cycles" (P. R. Boyle, ed.), Vol. I, pp. 143–157. Academic Press, London.

Yamamoto, M. (1982). Normal stages in the development of the cuttlefish, *Sepiella japonica* Sasaki. *Zoological Magazine (Dobutsugaku Zasshi)* 91, 146–157.

Yamamoto, M. (1985). Ontogeny of the visual system in the cuttlefish, *Sepiella japonica*. I. Morphological differentiation of the visual cell. *Journal of Comparative Neurology* 232, 347–361.

Yamamoto, M., Takasu, N. and Uragami, I. (1985). Ontogeny of the visual system in the cuttlefish, *Sepiella japonica*. II. Intramembrane particles, histofluorescence, and electrical responses in the developing retina. *Journal of Comparative Neurology* 232, 362–371.

Young, R. E. and Harman, R. F. (1985). Early life history stages of enoploteuthin squids (Cephalopoda: Teuthoidea: Enoploteuthidae) from Hawaiian waters. *Vie et Milieu* 35, 181–201.

Young, R. E., Harman, R. F. and Mangold, K. M. (1985). The common occurrence of oegopsid squid eggs in near-surface oceanic waters. *Pacific Science* 39, 359–366.

Parasitology of Marine Zooplankton

Jean Théodoridès

Laboratoire d'Evolution des Etres Organisés, Université P. & M. Curie, 105 Boulevard Raspail, 75006 Paris, France

ADVANCES IN MARINE BIOLOGY,
VOLUME 25 ISBN 0–12–026125–1

I. Introduction

Since its definition by Hensen (1887), plankton has been the subject of numerous studies from the point of view of its qualitative and quantitative composition, its fluctuations and migrations, its geographical distribution, its role in the productive cycle and its ultimate contribution to human nutrition. Although there have been reviews of diseases of marine animals in general (e.g. Kinne, 1980–1985), there has been no review of parasites of plankton. Existing articles are scattered in various specialized journals and there is no comprehensive work on a topic which, however important, does not seem to have aroused much interest either from planktonologists, or, with some rare exceptions, from parasitologists. This is surprising considering that the planktonic environment shows the most diversified examples of relations ranging from simple carrying of one organism by another (phoresy) to endoparasitism where the parasites' structure is greatly modified to suit the mode of life.

All the organisms comprising the marine zooplankton: Protista (Peridinians, Acantharia, Radiolaria etc.), Cnidaria (Medusae, Siphonophora), Ctenaria, Annelida, Crustacea (Amphipoda, Euphausiacea, Decapoda, Mysidacea), Mollusca and Tunicata (Salpida, Appendicularia) are parasitized by various organisms ranging from viruses and bacteria to Metazoa. These various forms of parasitism often show remarkable morphological and biological adaptations. Various examples will be given from different zoological groups to give a general biological impression of a very peculiar environment favourable for parasitism. It is also evident that parasites, especially the pathogenic species, must play a role in the fluctuations of planktonic populations, but this role remains to be determined.

II. Viruses and Bacteria

A. *Viruses*

Viral particles have been reported from free or parasitic dinoflagellates. Franca (1976) observed such intracytoplasmic particles in *Gyrodinium resplendens* (Hulburt) found on the west coast of Portugal, south of Lisbon; they measured either 35 nm in diameter and were then in "free" clusters, or 20 nm, when surrounded by a membrane. The author gives no indication of the exact nature of these inclusions which she compares with those observed in various marine algae. A related observation has been made by Sickogoad and Walker (1979) and concerns *Gymnodinium uberrimum*, a freshwater dinoflagellate from Lake Huron, in which these authors have reported the occurrence of viral particles of a pentagonal or hexagonal shape, with a diameter of about 385 nm and therefore ten times bigger than those reported by Franca and Soyer in marine hosts. The same authors have observed smaller particles with a diameter of 175 nm in a Chrysophycea (*Mallomonas* sp.). Mayer and Taylor (1979) found a virus which induced a lysis of the marine phytoflagellate *Micromonas pusilla* (Prasinophyceae). Soyer (1978) has found similar but intranuclear particles in a peridinian parasite of planktonic copepods, *Blastodinium* sp. The main cluster observed measured about 900 × 1000 μm and the average size of the virions was 30 nm. According to Soyer, the 20 nm particles surrounded by a membrane observed by Franca would correspond to sacs of trichocystoid filaments and not to viruses.

Pathogenic viruses certainly occur in marine planktonic invertebrates although there is not yet much published work on this topic. However, Bergoin *et al.* (1984) have recorded an infection by a *Chloriridovirus* (Iridoviridae) in populations of the cladoceran *Daphnia magna* in a brackish water environment (Camargue, France). The infected hosts take on a green iridescent colour, become inactive and die; the virus localized in the fat tissue showed a spherical outline in section and a diameter of 180 nm.

B. Bacteria

As is well known, plankton is very rich in Gram-negative bacteria called planktobacteria (Bauld and Staley, 1976; Sieburth, 1979). Some of these are free-living and have no close relations with the zooplankton except for the fact that they constitute a part of its food. Other bacteria (epibacteria) are attached to planktonic organisms or occur in their digestive tracts. Jones (1958) investigated the bacteria attached to the radiolarian *Castanidium longispinum* and found that there were about 50 000–100 000 bacteria/g of wet radiolarians.

Lear (1963) studied the chitinoclastic bacteria isolated from zooplankton samples in the Californian Pacific. Positive results were obtained with crushed radiolarians, chaetognaths, Crustacea (copepods, mysids, euphausiids, amphipods, decapods) taken with a standard net tow. With larger Crustacea (*Gnathophausia, Nematoscelis, Sergestes, Acanthephyra, Benthesicymus*), colonies containing from 1000 to 10 000 bacteria were obtained. Among the latter were Gram-negative asporogenous rods (69%), Gram-positive cocci (23.5%), Gram-positive rods (5.5%) and yeasts (2%).

Human pathogenic bacteria such as *Vibrio parahaemolyticus* are sometimes associated with zooplankton (Kaneko and Colwell, 1973). Simidu *et al.* (1971) studied the bacteria associated with coastal plankton in Japan and compared it with that of the neighbouring sea water. They established that 70% of heterotrophic bacteria from plankton samples (including zooplankton) were represented by the genera *Vibrio* and *Aeromonas* which constitute only 45% of the bacterial marine flora.

Epibiotic bacteria occur on planktonic ciliates (vorticellids). On the colonial species, *Zoothamnium pelagicum* they are found on the peduncular trunk and the zooids (Dragesco, 1948; M. Laval, 1968). These bacteria are mainly localized on the zooids and macrozooids; they may become very abundant and spread on the peduncles of the zooids and even on the trunk of the colony. They divide transversally, forming small chains of three to five individuals perpendicular to their support. The same author (M. Laval, 1970) reported from the same host bacteria localized in the macronucleus. A related species (*Zoothamnium alternans*) living in the benthic environment also bears epibiotic bacteria on the surface of its zooids, some of which are parasitized by bacteriophages (Fauré-Fremiet *et al.*, 1963).

A choanoflagellate, *Salpingoeca pelagica* epibiotic on *Z. pelagicum* is the host of intracytoplasmic bacteria localized in vacuoles attached to the mitochondria; it is not known whether they are symbionts or parasites (M. Laval, 1971). The acinetian *Tunicophrya sessilis* epibiotic on the tunicate *Pyrosoma elegans* is sometimes parasitized by filamentous bacteria which induce a deviation of its tentacles (Collin, 1912). Intracytoplasmic bacteria have been also reported from tintinnid ciliates, for instance in *Petalotricha ampulla* (M. Laval, 1972). In dinoflagellates, endocytoplasmic and endonuclear bacteria have been observed by Silva (1967) in *Cochlodinium heterolobatum* in the United States and by the same author (1978) in *Gymnodinium splendens* and *Glenodinium foliaceum* in Portugal. In the first host genus, the bacteria were endocytoplasmic whereas in the two others they were intranuclear.

In a more detailed work (Silva and Franca, 1985) the ultrastructural relationship between the bacteria of *Gyrodinium instriatum* and *Glenodinium foliaceum* and their hosts were studied and it seems that the hosts would benefit from the bacteria.

Cyanobacteria (formerly Cyanophyceae) are frequently encountered as symbionts of marine planktonic diatoms or benthic invertebrates (sponges, tunicates) (Taylor, 1973; Sieburth, 1979) but do not seem to occur frequently in zooplankton organisms. However Hall and Claus (1963) have mentioned "cyanellae" (symbiotic cyanobacteria) in the cryptomonad flagellate *Cyanophora paradoxa*.

III. Fungi

Only the "higher" fungi of previous classifications are presently placed in this kingdom, the "lower" ones being removed to the kingdom Protista (Corliss, 1986).

Some yeasts have been reported as parasites of planktonic cladoceran and copepod Crustacea. As Sieburth (1979) noticed, yeasts do not constitute a natural taxonomic entity and the marine species can belong to a variety of species of the Ascomycetes, Basidiomycetes and Deuteromycetes, i.e. the former "higher" fungi. Yeasts belonging to the genera *Monospora* and *Dispora* have been reported as pathogenic for the cladoceran *Penilia aviros-tris* (Rose *in* Trégouboff and Rose, 1957), the first of these genera having been observed previously by Chatton (1920) in the body cavity of a *Paracalanus* sp.

Ascomycetes of the genus *Metschnikowia* have been reported from *Cala-nus plumchrus*, a predominant copepod from the plankton of the Straits of Georgia and the brackish waters of the Nanaimo River and its estuary in British Columbia (Seki and Fulton, 1969). This genus is named after E. Metschnikoff who first isolated a species from *Daphnia magna*, a freshwater cladoceran. Seki and Fulton (1969) suggest that the numerous empty exoskeletons of the host found in the deep-sea would reflect a pathogenic effect on the parasite, whereas Sieburth (1979) thinks that the latter would be the consequence of normal moulting or the action of predatory ciliates. A similar infection by *Metschnikowia* has been described by Fize *et al.* (1970) in another copepod, *Eurytemora velox*, in the brackish waters of Camargue (France). The parasites fill the body of the host which they decimate, thus influencing the food chain.

IV. Protista

The kingdom Protista, as recently defined (Corliss, 1986), is composed of 18 assemblages divided into 45 phyla including organisms previously considered as plants (lower fungi, algae) or animals (Protozoa).

We shall only consider the organisms which have been reported as parasites (external or internal) or epibionts of planktonic animals. They belong to the Rhizopoda (amoebae), Mastigomycetes (lower fungi), Chloro-bionts (Chlorophyta), Polymastigotes (zooflagellates), Dinoflagellates (Peri-dinea, Syndinea, Ellobiopsidea), Ciliophora (ciliates), Sporozoa (= Apicom-plexa) (Gregarinia, Coccidia), Microsporidia and Paramyxea.

A. *Rhizopoda*

In the suborder Mastigogenina and the family Paramoebidae the genus

Janickina, created by Chatton (1952), includes two species; *J. pigmentifera* and *J. chaetognathi* that are parasites of the testicles and seminal vesicles of various chaetognaths of the genera *Spadella* and *Sagitta*. These parasites, discovered by Grassi (1881), were studied later by Janicki (1912a,b, 1928, 1932) and Hamon (1957) who refers to the occurrence of *J. pigmentifera* in chaetognaths from Villefranche-sur-mer and Algiers since Janicki's observations.

Hamon has studied the cytology of this amoeba and shown that it ingests the spermatocytes of the host. She also suggested that the infestation of the latter occurs during fertilization, the amoebae being attached to the spermatophore, thus enabling it to penetrate the body. More recently Hollande (1980) studied the "nebenkern" of *J. pigmentifera* on the ultrastructural level and showed that it is in fact a symbiotic organism related to the kinetoplastid flagellates, which he describes under the name *Perkinsiella amoebae* (Fig. 1).

Amoebae are not mentioned among the parasites of chaetognaths from Suruga Bay (Japan) by Nagasawa and Marumo (1979) and Shimazu (1979).

B. *Mastigomycetes*

Under this assemblage are grouped the former "lower fungi" with flagellated zoospores, such as the Phycomycetes, Oomycetes and Chytridiomycetes (Corliss, 1984). The latter are reported from planktonic dinoflagellates (Peridinians). Such is the case of *Amphicypellus elegans*, parasitizing species of *Ceratium* and *Peridinium* (Ingold, 1944, Canter, 1961), other species being found in other planktonic flagellates (Ingold, 1940; Canter, 1959). In the body cavity of planktonic copepods (*Paracalanus, Clausocalanus, Acartia*) Chatton (1920) observed a parasite consisting of ramified tubes which he considered as belonging to the genus *Ichthyosporidium*, described by Caullery & Mesnil (1905). There is an incipient taxonomic problem here, in that the same generic name has also been applied to certain fungi, but Sprague (1965, 1966) recommends its restriction to the microsporidians. Jepps (1937) observed an *Ichthyosporidium* in *Calanus finmarchicus* in the Clyde Sea area and showed that it very much resembled a chytridiomycete parasite of fishes, *Ichthyophonus hoferi*, suggesting that the latter could have been infected while feeding on parasitized copepods. Sproston (1944), in a detailed monograph on *I. hoferi*, discussed Jepp's (1937) hypothesis and did not find it acceptable as: (1) all the stages of the parasite's life cycle are found in the fishes; (2) the occurrence of the parasite in copepods is very rare.

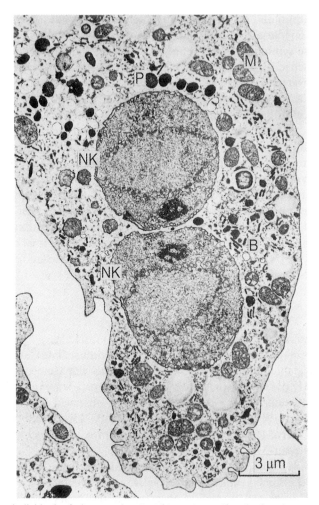

FIG. 1. An individual of the amoeba *Janickina pigmentifera* harbouring two parasitic kinetoplastid flagellates (NK) (*Perkinsiella amoebae*) (after Hollande, 1980).

C. *Chlorobionts*

Many unicellular organisms, formerly considered as unicellular algae (Dinophyceae, Chlorophyceae etc.), occur as symbionts in phyto- and zooplankton and they should be mentioned here although they are not parasites.

Lists have been given detailing their animal hosts (Mc Laughlin and Zahl, 1959, 1966; Norris, 1967; Droop, 1963; Dales, 1966; Taylor, 1973; Trench, 1987). These include zooplanktonic invertebrates such as Protista (radiolarians, acantharians, peridinians and other dinoflagellates, tintinnid ciliates, etc.) and Cnidaria (Scyphozoa, Siphonophora). The early workers designated these symbionts under the terms of "zooxanthellae", "zoochlorellae" and "cyanellae", if they were yellow to greenish-brown, pale to bright green, and bluish-green respectively. Droop (1963) showed the inadequacy of these colour criteria and Taylor (1971) demonstrated the lack of taxonomic significance of these various terms.

Among the so-called "zoochlorellae", several species of *Pedinomonas* (Prasinophyceae) occur in planktonic protista such as *P. noctulicae*, a symbiont of *Noctiluca miliaris* in Southeast Asia (Sweeney, 1976), whereas *P. symbiotica* occurs in the radiolarian *Thalassolampe margarodes* in the French Mediterranean Sea (M. Cachon and Caram, 1979). Euglenians such as *Protoeuglena noctilucae* have been found in *Noctiluca* in India (Subrahmanian, 1954). The "flagellate" reported by Berkeley (1930) as symbiotic with the ctenophore *Beroë abyssicola* would correspond to animal cell inclusions (Taylor, 1973). The symbiotic peridinians ("zooxanthellae") will be mentioned in Section E on the dinoflagellates.

D. *Polymastigotes*

This assemblage includes the former "higher zooflagellates", some of which are parasites of zooplanktonic hosts. Such is the case of the bodonid *Trypanophis grobbeni*, a parasite of siphonophores (*Macrophyes, Halistema, Cucubalus*), described by Poche (1903), and observed later by Keysselitz (1904), Floyd (1916) and Duboscq and Rose (1933) who tried to elucidate its life cycle. Carré (1968) observed a related form in *Sphaeronectes* spp. in Villefranche-sur-Mer.

Meanwhile, Duboscq and Rose (1926, 1927) had discovered another species of the same genus (*T. major*) in *Abylopsis pentagona*. These authors, as well as Rose (1939, 1947), observed in the hosts, beside the typical "flagellates" stages, the occurrence of "gregarinian" stages which they supposed to belong to the same parasite. Rose (1947) added to this rather curious life cycle two other stages ("mycelians" and "spirometoids") which were most probably artefacts or the result of a secondary contamination of the hosts.

More recently Cachon *et al.* (1972) studied the ultrastructure of *T. grobbeni* and showed that it belonged to the Bodonidae. According to them, the "gregarinian" stages would originate from another parasite belonging to the Sporozoa (Apicomplexa). A further ultrastructural study of the same parasite (Brugerolle and Charnier, 1981) suggested that it would be related to *Spiromonas*, to some dinoflagellates and to Sporozoa but the authors are still not sure whether the "gregarinian" stages belong to the life cycle of the same parasite. However the occurrence of a third species of *Trypanophis* in the digestive tracts of chaetognaths (*Sagitta*) (Hovasse, 1924b; Rose and Hamon, 1950; Hamon, 1951a), in the life cycle of which the "gregarinian" stages are also present, suggests that they belong to the cycle of *Trypanophis*. This would support the idea that the Apicomplexa are phylogenetically related to Polymastigotes which was suggested earlier by Chatton and Biecheler (1934, 1936).

E. *Dinoflagellata*

1. Symbionts

Dinoflagellates are "the most ubiquitous of all marine algal symbionts" (Taylor, 1973) often present in radiolarians (Anderson, 1976a,b; Trench, 1987) and also in planktonic foraminiferans such as *Globigerinoides* (Lee *et al.* 1965; Anderson and Bé, 1976) which harbour a *Symbiodinium* similar to the one found in the scyphozoan *Cassiopeia* sp. described by Freudenthal (1962). *Symbiodinium* has a typical peridinian cycle including vegetative cells, zoosporangia and zoospores; the relationship with its host was discovered by Balderston and Claus (1969).

Hollande & Carré (1974) studied the ultrastructure of the "xanthellae" from various radiolarians (*Thalassophysa, Thalassicolla, Collozoum, Collosphaera, Sphaerozoum*), an acantharian (*Acanthometra*) and the siphonophore *Velella velella* found in the plankton off Villefranche-sur-mer. They observed the peridinian *Endodinium nutricola* in the radiolarians, an *Endodinium* species and an undetermined "xanthella" in *Acanthometra* and *Endodinium chattoni* (described by Hovasse, 1923a, 1924a) in *Velella*. These authors note that many free-living planktonic *Gymnodinium* may, in fact, be flagellate stages from "xanthellae".

2. Parasites

Edouard Chatton (1883–1947) initiated since 1912 studies on parasitic peridinians found in various invertebrates (Protista, Cnidaria, Polychaeta, Crustacea, Tunicata) as well as in vertebrates (fishes and their eggs), most of them being planktonic. The same author (Chatton, 1952) gave, in a posthumous work, a useful classification of these parasites according to their localization in the host. Among the intestinal parasites of Metazoa is the genus *Blastodinium* found in copepods. Among the ectoparasites we have the genera *Chytriodinium* (in copepod eggs), *Apodinium, Parapodinium* (in

Appendicularia), *Oodinium* (in various invertebrates and fishes) and *Proto-odinium* (in cnidarians); other peridinians seen later are parasitic in other Protista.

The work of Chatton has, since 1964, been carried on by J. and M. Cachon and various other protistologists. A catalogue of parasitic and symbiotic species has been compiled by Sournia *et al.* (1975) and a monograph by Cachon and Cachon (1987). J. Cachon's initial monograph (1964) is entirely devoted to the peridinians that are parasitic in other Protista (radiolarians, acantharians, peridinians, etc.) from the zooplankton off Algiers and Ville-franche-sur-mer and concerns mainly their cytology and life cycles.

Among them one finds the Duboscquellidae parasitic in Ciliophora (tintinnids) of which Cachon (1964) described several species of the genera *Duboscquella, Dogielodinium* and *Keppenodinium*, the last two being parasites of acantharians and radiolarians.

In addition, Cachon created the family Amoebophryidae, describing the life cycle of six species of *Amoebophrya* parasitic in peridinians, heliozoa and tintinnids. In his monograph, he insisted on the morphological modifications of these parasites which made it difficult to establish their affinities with free-living species.

Cachon also noted that Chatton and Hovasse used to link the parasitic peridinians with phototrophic free-living species and recalled the successive steps from an osmo-phototrophic physiology (*Protoodinium, Blastodinium*) to an osmotrophic one and showed that some species are osmo-phagotro-phic. Cachon and Cachon (1969b) showed that the first part of the *Amoe-bophrya* life cycle (trophonts) is characterized by an intracellular parasitism, whereas the second part of the cycle (sporonts) is free-living and phagotro-phic.

The same authors (1974a, 1977) revised the family Apodinidae created by Chatton (1920). In four species, they describe the fixation and absorption systems noting an astonishing structural duality (two rhizoids, two condylae and two parallel elements which constitute the peduncle). They also (1974a, 1977) studied mitosis in free-living and parasitic *Oodinium* and showed that, during sporogenesis, all the transitions between the first and the second are observed.

The same structural comparison was made by them (1974b) in the stomato-pharyngeal system of the free-living genus *Kofoidinium* compared with other free-living or parasitic genera such as *Protoodinium, Duboscquella, Amoebophrya*.

The latest classification was given by the same authors (Cachon and Cachon, 1987):

Dinococcida (semi-parasitic on Siphonophora, semi-free: *Cystodinium, Sty-lodinium, Cystodinedria*).

True parasites:

I. Blastodinida (entirely extracellular)
1. Blastodinidae (*Blastodinium*) on copepods.
2. Protoodinidae (*Protoodinium*) on Cnidaria.
3. Apodinidae (*Apodinium*) on appendicularians.
4. Haplozoonidae (*Haplozoon*) on polychaetes.
5. Oodinidae (*Oodinium, Oodinioides, Amyloodinium, Crepidoodinium, Piscinoodinium*) on acantharians, polychaetes, appendicularians, fishes).
6. Chytriodinidae (*Chytriodinium, Myxodinium*) in crustacean eggs, *Halosphaera*.
7. Diplomorphidae (*Cachonella*) on Siphonophora.

II. Duboscquodinida (partly intra-, partly extracellular)
1. Amoebophryidae (*Amoebophrya*) in various Protista.
2. Duboscquellidae (*Duboscquella, Duboscquodinium, Keppenodinium*) in various Protista.
3. Sphaeriparidae (*Sphaeripara, Atlanticellodinium*) in Protista and appendicularians.

III. Syndinida (entirely intracellular)
1. Syndinidae (*Syndinium, Solenodinium, Synhemidinium, Ichthyodinium, Cochlosyndinium, Merodinium, Haematodinium, Trypanodinium*) in Protista, crustacean and fishes eggs.

We must now examine some of the members of this class mentioned here in relation to their various hosts, their life cycles being summarized if known.

(a) *Peridinians from Protista*

(i) *Radiolarians as hosts.* The species of *Syndinium* are found in the intracapsular endoplasm of the hosts (*Collozoum, Sphaerozoum, Myxosphaera, Aulacantha*, etc.) (Hollande, 1974). The colonies of *Collozoum*, parasitized by the peridinians, carry on their development, but if this is not synchronous in the host and the parasite, the host is destroyed by the liberation of the parasite's anisospores, which induces a lysis of the radiolarian jelly.

The genus *Solenodinium* is found only in the family Thalassicollidae; it is an endonuclear parasite which perforates the host's nuclear membrane (Fig. 2(A)) (Hollande and Enjumet, 1953; Hollande, 1974).

Caryotoma bernardi is a parasite of *Thalassicolla* and its life cycle is as follows: penetration of a gymnodinian spore into the endocapsular endoplasm of the host, growth of the spore by osmotrophy which changes it into a large trophozoite producing zoospores which are liberated in sea water after the lysis of the host's capsular envelope (Hollande and Corbel, 1982) (Fig. 2(B)).

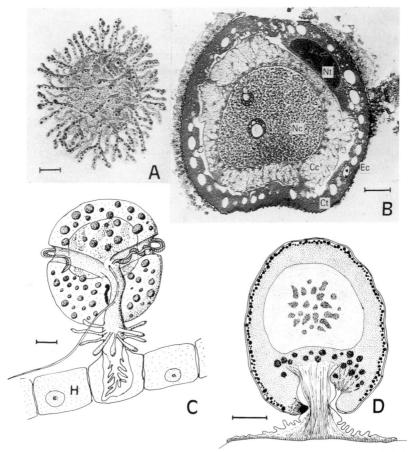

FIG. 2. Peridinians from various zooplanktonic hosts. (A) *Thalassicolla spumida* (radiolarian) parasitized by *Solenodinium fallax* of which the plasmodial ramifications perforate the host's nuclear membrane. Scale bar = 50 μm. (B) Trophozoite of *Caryotoma bernardi* in the endoplasm (Ct) of *Thalassicolla nucleata* (Cc: cytoplasm of the parasite; Ec: capsular envelope of the host; Nt: its nucleus; Nc: parasite's nucleus. Scale bar = 50 μm. (C) *Protoodinium chattoni* attached to the medusa *Podocoryne minima* (H). Scale bar = 10 μm. (D) *Oodinium fritillariae* attached to the appendicularian *Fritillaria*. Scale bar = 50 μm. [(A) after Hollande and Enjumet, 1953, (B) after Hollande and Corbel, 1982, (C) after Cachon and Cachon, 1971a, (D) after Cachon and Cachon, 1971b].

Atlanticellodinium tregouboffi is a parasite of *Planktonetta atlantica* (phaeodarian) and has been studied by Cachon and Cachon (1965) who place it near the genus *Sphaeripara*, parasitic in appendicularians.

Among the Amoebophryidae we must mention *Amoebophrya sticholonchae* which is parasitic in *Sticholonche zanclea* and was studied by Cachon (1964).

(ii) *Acantharians as hosts*. These planktonic Protista, related to the radiolarians, are parasitized by various peridinians. Among them are several species of *Amoebophrya*, such as *A. acanthometrae*, which destroys the nuclei of various species of *Acanthometra* and can itself be parasitized by another hyperparasitic *Amoebophrya*. In the genera *Dogielodinium* and *Keppenodinium*, parasites of acantharians are also found.

(iii) *Foraminifera as hosts*. Planktonic Foraminifera harbour various peridinians fixed to their gelatinous capsule or to their spines. Such is the case in *Hastigerina pelagica* on which a *Dissodinium* is found (Alldredge and Jones, 1973). These authors have shown that the number of parasites is in direct relation to the size of the host and that they must be considered more as commensals than harmful parasites.

In some cases the association between peridinians and Foraminifera is a symbiotic one as in *Globigerinoides sacculifer* (Anderson and Bé, 1976).

(iv) *Peridinians as hosts*. Some peridinians parasitize their free-living relatives, as in some species of *Amoebophrya* (Chatton and Biecheler, 1935). *A. grassei* is a hyperparasite of *Oodinium poucheti* which is itself parasitic in appendicularians; another *Amoebophrya* parasitizes *Gonyaulax catenella*, a toxic peridinian (Taylor, 1968). As regards the relations between these parasites and their hosts, they start as simple ectoparasites later to become endoparasites. Cachon (1964) very rightly compared the infestation of a *Gymnodinium* by an *Amoebophrya* to that of a crab by a sacculine.

The genus *Coccidinium*, of uncertain affinities (Cachon, 1964), is parasitic in the nucleus and cytoplasm of various peridinians (Chatton and Biecheler, 1936). The young trophozoite lies near the nucleus which it depresses before invading and destroying it. There are two kinds of trophozoites: in the first there is much starchy matter, whereas in the second this is rare or nonexistent. Each of these trophozoites will divide to give flagellate dinospores with a "helmet".

Various species of *Duboscquella* are parasitic in various peridinians (gymnodinians, Peridinidae, *Noctiluca, Leptodiscus*, etc.)

(v) *Ciliophora as hosts*. *Amoebophrya rosei* is a parasite of several Apostoma of the family Foettingeridae which are parasites of Siphonophora and Chaetognaths.

The Duboscquellidae are parasites of tintinnids (Fig. 3). For instance, *Duboscquella tintinnicola* is found in the genera *Codonella, Tintinnopsis, Cytarocylis, Tintinnus* and *Rhabdonella*. This species appears as an ovoid mass which reproduces itself by fission giving gametocytes and later biflagellate gametes. The life cycle of other species of *Duboscquella* are described by

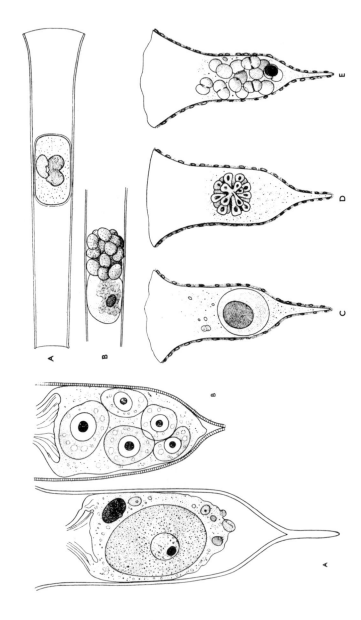

FIG. 3. Peridinians parasitic in Tintinnids. Left: *Duboscquella tintinnicola* in *Cytarrocyclis ehrenbergi*. (A) Single individual, (B) Five individuals (after Collin *in* Chatton, 1952). Right (A, B) *Duboscquodinium collini* in *Tintinnus fraknoii*. (C–E) *Duboscquodinium kofoidi* in *Codonella campanula* [(C) trophozoite, (D) rosette of the gametocyte, (E) gymnospores] (after Grassé *in* Chatton, 1952).

Cachon (1964); among them, *D. caryophaga* is parasitic in *Strombidium*, *Strombilidium* and *Prorodon D. cachoni* in *Eutintinnus* (Wayne Coats, 1988).

The related genus *Duboscquodinium* is also a parasite of tintinnids; its trophozoites become trophocytes and gametocytes giving 16 gymnospores (Fig. 3).

(b) Peridinians from Cnidaria

(i) *Anthomedusa as hosts*. Cachon and Cachon (1971a) studied the relations between *Protoodinium chattoni* and the medusa *Podocoryne minima* on the manubrium of which the peridinian is fixed by a peduncule which works like an absorbing sucking device (phagotrophy) (Fig. 2(C)).

(ii) *Siphonophora as hosts*. Rose and Cachon (1951, 1952a,b) and Cachon (1953) studied *Cachonella paradoxa*, an ectoparasite of various Siphonophora (*Chelophyes, Abylopsis, Forskalia* etc.) in the internal part of the swimming bells or bracts. The trophozoites are attached to the host by numerous rhizoids and later reach the digestive cavity, where they undergo a transformation giving stages with finger-like appendages. Later they are released in seawater and become sporonts from which will emerge flagellate zoospores (Fig. 4(A)(B)).

Cachon *et al.* (1965) investigated the life cycle of *Stylodinium gastrophilum*. This parasite of the gastrozoites of *Abylopsis, Chelophyes, Sulculeolaria* previously studied by Cachon (1964) has trophonts fixed by a peduncle.

(c) Peridinians from Polychaeta

Eleven species of the genus *Haplozoon* are known, all parasitic in polychaetes of which some are planktonic. The young stages are attached to the intestinal epithelial cells of the worms by means of a stylet and retractile filaments (Chatton, 1920). Then the proximal cells give rise to distal ones which develop into trophocytes and later into sporocytes which produce flagellate dinospores.

Some species of *Oodinium* are parasitic in alciopids such as *O. dogieli* discovered by Dogiel (1910a) and again reported by Cachon and Cachon (1971b) in Spionidae.

(d) Peridinians from Crustacea

It is precisely in planktonic Crustacea (copepods) that the first parasitic peridinians were discovered; they belong to the genus *Blastodinium* and were localized in the digestive tract (Chatton, 1906). Many other papers by the same author followed, especially his important monograph (Chatton, 1920).

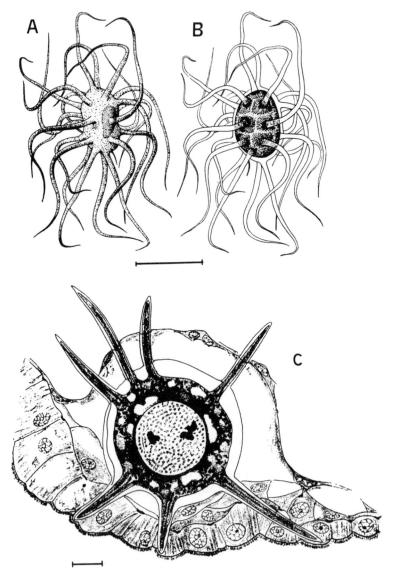

FIG. 4. (A, B) *Cachonella paradoxa*, a peridinian parasite of various Siphonophora. (A) Individual from the host's gastrozoid, (B) Individual in seawater where the condensing of the cytoplasm into a central mass is very noticeable. Scale bar = 100 µm (after Cachon, 1953). (C) *Actinodinium apsteini*, a parasite of the body cavity of the copepod *Acartia clausi*, was once considered a peridinian but it probably belongs to another taxon. Scale bar = 100 µm (after Chatton and Hovasse, 1937).

He also showed that the consequences of this parasitism for the hosts were as follows:

(1) an indirect parasitic castration;
(2) the inhibition of the male host's last moult or its death during this moult and,
(3) the decrease of the average size of the parasitized hosts.

Sewell (1951) devoted an important monograph to the parasitic peridinians of the Arabian Sea. They belong to the two genera *Blastodinium* and *Syndinium*. The effects on their hosts have already been studied by Hovasse (1923b). In female individuals of *Clausocalanus* and *Paracalanus*, modifications occur in the development of the fifth pair of legs and in *Undinula* the same modifications occur in the genital segment, where a failure of the fusion between the first and the second abdominal segments was observed. There is, however, no parasitic castration, the ovaries and eggs being present in parasitized females.

Among other works on peridinians of Crustacea must be mentioned that of Cachon and Cachon (1968) in which is described the cytology and the life cycle of three species of *Chytriodinium* parasitic in eggs of copepods and euphausiids.

Drebes (1969, 1972) described *Dissodinium pseudocalani*, a parasite of the eggs of *Pseudocalanus*, of which the life cycle is as follows: the primary cyst, which is free-living in the plankton, has a multinucleated cytoplasm dividing into 16 cells each developing into as many secondary cysts; from these dinospores develop which attach themselves to the host's egg, absorbing its contents and becoming primary cysts. The same author (Drebes, 1978) studied a related species, *D. pseudolunula*, and Soyer (1974) a *Syndinium* sp. from the body cavity of copepods.

Some other parasites of Crustacea described as being peridinians are considered dubious by specialists of this group. Such is the case of *Actinodinium apsteini* (Fig. 4(C)), a parasite of copepods (*Acartia*) found by Chatton and Hovasse (1937). In the same hosts, Chatton (1920, 1927, 1940) studied *Paradinium poucheti*, later re-examined by Cachon *et al.* (1968) and Chatton and Soyer (1973), which seems related to the Mycetozoa (Myxomycetes).

(e) Peridinians from Tunicata

These parasites are of special biological interest since their association with their hosts ranges from phoresy to ectoparasitism and endoparasitism. They chiefly occur in appendicularians and sometimes in salpids. As phoretic species one may mention *Filodinium hovassei*, attached to the cuticle of

Oikopleura, which has the characteristic features of free peridinians and well developed rhizopodial system allowing it to crawl over the host's body.

The ectoparasites are found in the genera *Oodinium* and *Apodinium*. The former includes species studied as early as 1912 by Chatton in Mediterranean hosts. The rather complex structure and ultrastructure of *Oodinium* was later studied by Cachon *et al.* (1970) who described the pusules, organelles that may have an excretory function. Cachon and Cachon (1971a) worked out the ultrastructure of the attachment stalk and showed that the food intake of *Oodinium fritillariae* (parasitic on *Fritillaria*) takes place by phagotrophy; they also showed that at the level of the attachment device, the plasmalemma send many finger-like invaginations into the stalk (Fig. 2(D)).

Oodinium amylaceum is found on the gills of salpids. The genus *Apodinium* includes several species, all ectoparasitic on appendicularians, and showing some site-specificity on the host: *A. mycetoides* attaches near the spiracles of *Fritillaria*, *A. rhizophorum* on the tail of *Oikopleura* and *A. chattoni* on that of *Fritillaria*, whereas *A. zygorhizum* is localized near the mouth (Cachon and Cachon, 1973).

Sphaeripara (= *Neresheimeria*) *catenata* was discovered in 1883 and studied by Chatton (1920). It is a parasite of the gonads of *Fritillaria* and its zoological position has for a long time been a mystery. Its segmentation by membranous diaphragms had led some naturalists to consider it as a Mesozoan. Cachon and Cachon (1964, 1966) showed it to be unquestionably a peridinian while studying the ultrastructure of its cuticle and its development in the formation of the diaphragms and the relation between the parasite and the host's cytoplasm. Some authors have claimed that *S. catenata* induces parasitic castration in its host, since it is found in the gonad. In fact castration only occurs if the infestation of this organ has taken place before its full development. *Syndinium oikopleurae* parasitizes the gonads of *Oikopleura* (Fenaux, 1963; Hollande, 1974).

(f) Peridinians from fishes

While some peridinians are parasitic in the planktonic eggs of marine fishes, others (*Amylodinium, Oodinium, Piscinoodinium, Crepidoodinium*) have been reported from adults (Brown, 1931, 1934; Brown and Hovasse, 1946; Lom, 1976, 1981; Lom and Lawler, 1973). Among the former, *Ichthyodinium chabelardi* is a parasite of the eggs of *Sardina pilchardus* (Hollande and Cachon, 1952, 1953). Reproduction occurs by schizogony, the young stages being spherical with a diameter less than 20 µm; they then grow to become giant plasmodes reaching a diameter of 100 µm, the latter multiplying by transverse fission to give secondary triangular-shaped schizonts which become zoospores. The rate of infestation of the eggs varies from 30 to 80%

and there is a significant mortality in the parasitized fry, following the bursting of the yolk sac.

To conclude, one must recognize the importance of the peridinians as parasites of marine zooplankton, where they are found in all zoological groups from the Protozoa to the vertebrates. In their association with their hosts, they show all the steps between phoresy, ecto- and endoparasitism. The ectoparasitic species have special devices—stalks, rhizoids, sucking organs, etc.—allowing them to feed by osmotrophy or phagotrophy on the host's tissues.

Some genera, such as *Sphaeripara*, stand on the border between ecto- and endoparasitism, only a part of the parasite being buried in the host. In the case of the ectoparasites, the effect on the host is generally negligible but, as is generally the case with all parasites, the endoparasites are much more noxious. Such is the case with *Solenodinium fallax*, a parasite of *Thalassicolla* (radiolarian) (Fig. 2(A)), the nucleus and later the cytoplasm of which are totally invaded by the plasmodial tubes of the peridinian. Similarly *Sphaeripara* can induce castration of its host, Reciprocally the host's environment may have a definite influence on the life cycle of the peridinians. This occurs with *Apodinium, Chytriodinium* and *Oodinium* in which one observes a multiple division as soon as the trophic contact with the host is interrupted.

F. *Ellobiopsidea*

These curious organisms defined by Grassé (1952b) are all parasitic on Crustacea, mainly on planktonic species of Decapoda, Mysidacea, Euphausiidae and Amphipoda, and on Polychaeta (Caullery, 1910; Coutière, 1911; Fage, 1936; Nouvel, 1941; Boschma, 1949). They look like fungi, each individual consisting of one or several tubes generally ramified and nearly always septate. In most cases the proximal segment or trophomere sends a sucking device into the body of the host. Hovasse (1926) showed the life cycle of *Parallobiopsis coutieri*, a parasite of *Nebalia*, to be as follows: the young attached organism becomes ovoid or cylindrical then an attachment apparatus and a sucking device develop. The nuclei multiply and, when the parasite has reached 50 μm diameter, it becomes transversally septate with two segments: the proximal or *trophomere*, and the distal or *gonomere*. After some time, the nuclei of the gonomere lift its membrane and bud outside it; they then acquire a flagellum and are liberated as spherical spores that fix themselves to a new host.

This occurrence of flagellated spores has led some authors (Galt and Whisler, 1970; Hovasse, 1974; Loeblich, 1976) to consider the Ellobiopsidea as belonging to the Dinoflagellata but this is not accepted by the specialists of this group (Cachon and Cachon, 1987) who think they might belong to a separate class.

However in the present paper we shall follow the classification of Protista given by Corliss (1984) and consider the Ellobiopsidea as associated with dinoflagellates. There are less than 10 species described.

The genus *Ellobiocystis* is found on the mouthparts and maxillipedes of various species of planktonic Decapods such as *Acanthephyra, Pasiphaea* and *Sergestes.*

Ellobiopsis (Fig. 5) is a parasite of copepods, *Thalassomyces* of mysids, euphausiids, amphipods and decapods (Caullery, 1910; Coutière, 1911; Hovasse, 1952; Jones, 1964; Kane, 1964; Vader, 1973a,b; Vader and Kane, 1968).

These organisms are not harmless epibionts, as was first thought when they were discovered by Sars (1868), but real endoparasites in which the

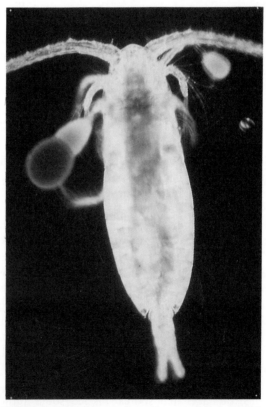

FIG. 5. Two *Ellobiopsis* sp. attached to a calanid copepod from the Subantarctic (original photograph by J. B. Crumeyrolle).

sucking devices penetrate deeply into the body of their hosts. In *Thalassomyces marsupii*, a parasite of *Parathemisto* (Amphipoda, Hyperiidae) the sucking device may reach vital organs of the host such as the nerve ganglia (Kane, 1964), whereas in *Thalassomyces fagei*, a parasite of euphausiids, it penetrates the ovary of the host (Boschma, 1948; Ramirez and Dato, 1989).

Mauchline (1966) has shown that this species occurs more frequently in the large males and the small females of the host (*Thysanoessa raschi*) than in the small males or large females. According to the same author, the seminal vesicles of males parasitized by *T. fagei* are less developed than in non-parasitized males. Several of the parasitized females had small ovaries and did not produce eggs when mature.

T. fagei has been found on other euphausiids, e.g. *Meganyctiphanes norvegica* (Bhaldraithe, 1973), *Thysanoessa inermis* (Lindley, 1977). Fage (1936) had already shown that in the adult female of *Gnathophausia zoea* (a mysid) parasitized by *Thalassomyces fasciatus*, the sternal plate remained small. Similarly the fifth pereiopod is half the length of that in the normal female. *Thalassomyces albatrossi* parasitizes *Stylomysis* (Wing, 1975) while the genus *Amallocystis* is also found on mysids (Nouvel, 1941; Nouvel and Hoenigman, 1955; Hoenigman, 1960, 1965).

G. *Ciliophora*

This assemblage includes many Protista associated with zooplanktonic invertebrates, the association ranging from epibiontism to endoparasitism. The classification used here is that of Corliss (1979). Class Kinetofragmophora, Subclass Hypostomata. In the order Microthoracida, the family Conchophryidae includes the genus *Conchophrys* with one species, *C. davidoffi*, a commensal of pyrosomes (*Pyrosoma giganteum*) which lives in the buccal siphons (Chatton, 1911a).

In the order Rhynchodida, one finds *Heterocoma hyperparasitica*, a parasite of *Actinobranchium salparum*, which is a suctorian living on the pericoronal arches of *Salpa democratica* (Chatton and Lwoff, 1939).

In the order Apostomatida, created and studied in detail by Chatton and Lwoff (1935), the family Foettingeridae includes several endoparasites of planktonic invertebrates. *Vampyrophrya pelagica* is a parasite of copepods such as *Acartia, Paracalanus* and *Clausocalanus*; the phoronts of the ciliate hatch in a damaged host, develop in its body, then emerge and encyst in the sea where they undergo palintomy, the tomites reproducing the phoronts. The other genera of Foettingeridae are: *Traumatiophora* (parasite of *Calanus*), *Pericaryon, Perezella* and *Metaphrya*. *Pericaryon cesticola*, studied by Chatton (1911a), lives in the gastrovascular canal of the ctenophore *Cestus veneris. Perezella pelagica* and *Uronema rabaudi*, parasites in the body cavity of copepods (*Clausia, Acartia, Clausocalanus*) and described by Cépède (1910), are considered by Chatton and Lwoff (1935) as related to the Apostomatida. *Metaphrya sagittae*, described from Japan by Ikeda (1917), is a coelomic parasite of chaetognaths (*Sagitta* spp.). Recently it has been frequently reported from various hosts in Japan (Shimazu, 1979; Nagasawa and Marumo, 1979). It had previously been found in *Sagitta minima* in the Moroccan Atlantic by Furnestin (1957) and in *Sagitta elegans* in Canada by Weinstein (1972).

Several other unnamed apostomes have been reported by Rose (1936) in Algerian siphonophores (*Abylopsis, Galeolaria*) and copepods (*Spinocalanus, Aetideus, Chiridius, Drepanopsis, Euchirella, Euchaeta, Phaenna, Scolecithircella, Ammallothrix, Haloptilus, Augaptilus, Euaugaptilus, Onchocalanus, Pleuromamma*), by Sewell (1951) from copepods of the Arabian Sea (*Nannocalanus, Paraeuchaeta, Scottocalanus, Amallothrix, Lophotrix*), and by Lindley (1978) from North Atlantic and North Sea euphausiids. *Jeppsia(= Chatonella) calani* is an ectoparasite of *Calanus finmarchicus* (Jepps, 1937).

The subclass Suctoria (= Acineta or Tentaculifera) includes many Ciliophora phoretic or commensal on freshwater or marine invertebrates. Among the latter are several planktonic hosts. Rose (*in* Trégouboff and Rose, 1957) has given a list of the main genera: *Acineta* (epibiont on copepods and salps), *Paracineta, Hallezia* (copepods), *Ephelota* (copepods, pteropods, pyrosomes), *Actinobranchium* (salps), *Thecacineta* (ostracods), *Hypocoma* (on other Ciliophora: *Zoothamnium, Ephelota, Acineta*) *Tokophrya, Corynophrya* (copepods). Sewell (1951) has described several species of the genera *Acineta, Paracineta* and *Hallezia* as epibionts of copepods of the Arabian Sea. Sherman and Schaner (1965) studied the association of *Paracineta* sp. with the copepods *Metridia lucens* and *Metridia longa* in the Gulf of Maine. There is a certain specificity in this association since the suctorian i.e. *Paracineta* has neither been found on other copepods nor on other planktonic invertebrates.

The authors identified three categories of *Paracineta* incidence on *Metridia* based on the number of suctorians present per copepod: slight (1–50), moderate (50–100) and heavy (100–250). The heaviest incidence on female hosts is explained by their large size offering a larger substratum. There are no significant differences in vertical distribution of *Paracineta* incidence on *Metridia* among the different depths samples (0, 10, 30, 60 m) and no harmful effect of the parasite was observed.

Nicol (1984) reported the occurrence of an *Ephelota* sp. on the euphausiid *Meganyctiphanes norvegica* in the Bay of Fundy. The parts of the host most frequently infested were the pleopods, the ventral abdomen and the eyestalk area. Between 8 and 82% of the hosts from surface swarms were infested, whereas the infestation was less than 2% in samples taken at offshore depths.

This is probably due to the fact that the euphausiids in the surface swarms were older. *Ephelota gemmipara* lives on the pteropod *Euclio*.

Among the Dendrosomatidae, *Actinobranchium*(= *Trichophrya*) *salparum* lives on the pericoronary arches and sometimes in the inhalant siphon of various salps (chiefly *Thalia democratica*). *Brachyosoma scottocalani* lives on the copepod *Scottocalanus* in the Arabian Sea (Sewell, 1951). In the Ephelotidae, *Tunicophrya sessilis* lives on the tunica of *Pyrosoma elegans* and several species of *Corynophrya* (Thecacinetidae) occur on copepods.

In the suborder Mobilina of the order Peritrichida, *Trichodina* are frequently found in planktonic invertebrates (the present author has frequently observed a *Trichodina* sp. in the pallial cavity of the mollusc *Clio cuspidata* in the plankton off Villefranche-sur-Mer).

In the suborder Licnophorina of the order Heterotrichida, *Licnophora luidiae* has been described off Algiers by Hamon (1954), as being ectoparasitic on planktonic larvae (bipinnaria) of the asterid *Luidia sarsi*. This ciliate has a sucker to aid attachment to the host.

H. *Sporozoa (Apicomplexa)*

In this group are found exclusively parasitic Protista classified in three subclasses: Gregarinia, Coccidia and Hematozoa of which only members of the first two occur in marine zooplanktonic hosts.

1. Gregarinia

Gregarines are found in various planktonic invertebrates (Polychaeta, Chaetognatha, Crustacea, Mollusca, Tunicata) and are localized in the digestive tract or body cavity of their hosts.

For an easier presentation they will be considered here in relation to the zoological position of the hosts.

(a) *Gregarines from Polychaeta*

(i) *Coelomic parasites.* Mingazzini (1891, 1893) was the first to mention trophozoites of a gregarine named *Lobianchella beloneides* in the body cavity of an alciopid (*Alciopa* sp.). More recently Théodoridès and Carré (1969) again found this parasite in *Naiades cantrainii* from Villefranche-sur-Mer and showed that it belonged to the genus *Gonospora* including parasites from polychaetes and molluscs (Fig. 6(A)). A study of its ultrastructure was given later (Corbel *et al.*, 1979) (Fig. 6(B, C)).

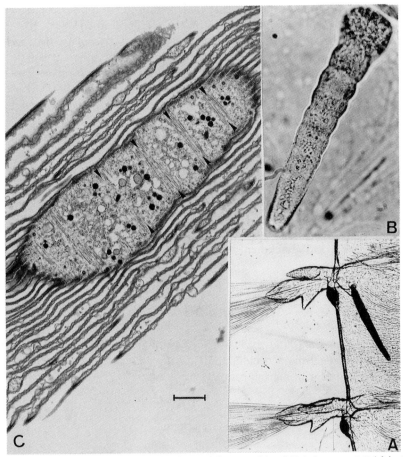

FIG. 6. *Gonospora beloneides*, a coelomic gregarine parasitic of *Naiades cantrainii* (alciopid polychaete). (A) Localization of a parasite in the host. (B) Trophozoite observed *in vivo*. (C) Ultrastructure of a young trophozoite showing the bundles of fibrils (arrows). Original scale bar = 1 μm. (A) After Théodoridès and Carré, 1969. (B) After Corbel *et al.*, 1979c.

(ii) *Intestinal parasites.* The last named authors recalled the observations of Greeff (1885) who described in the intestine of *Rhynchonerella fulgens* a nematoid and annulated gregarine called *Gregarina annulata*. Since this genus is restricted to parasites of insects, this species was placed by Labbé (1899) in the genus *Polyrhabdina*. Actually it should be placed in the genus *Lecudina* from which Théodoridès and Carré (1969) reported, from *Rhynchonerella petersii*, a related but not annulated species. Corbel *et al.* (1979) described *Lecudina danielae* from the digestive tract of *Vanadis crystallina* in Villefranche-sur-Mer.

(b) Gregarines from chaetognaths

(i) *Coelomic parasites.* In the body cavity of *Sagitta punctata* from Ville-franche-sur-Mer, the present author observed elongated cysts belonging most probably to an undescribed gregarine (Fig. 7(B)).

(ii) *Intestinal parasites.* Mingazzini (1891, 1893) reported *Lankesteria leuckarti* from *Sagitta* sp. in the Gulf of Naples. Nearly a century later this gregarine was rediscovered in several species of *Sagitta* and in *Krohnitta pacifica* in Suruga Bay (Japan) by Nagasawa and Marumo (1979). It was also studied from the same material by Shimazu (1979) who was uncertain of its taxonomic status and reported the occurrence of two other gregarines from the same hosts. The original discovery at Naples was at first included in *Lecudina* but was removed to *Lankesteria* by Labbé (1899) on account of its non-septate features. The morphology of Shimazu's two other gregarines resembles that of another intestinal gregarine described from *Sagitta lyra* by Hamon (1951b) under the name *Tricystis planctonis*. It seems, in fact, to belong to the genus *Sycia*. Additional information on gregarines from chaetognaths is available in the monographs of Furnestin (1957) and Weinstein (1972).

(c) Gregarines from Crustacea

Three families of Eugregarines occur in planktonic Crustacea: Cephaloido-phoridae, Cephalobidae, Ganymedidae. These only include intestinal para-sites.

(i) *Cephaloidophoridae.* Over 10 species of *Cephaloidophora* have been recorded from hyperiid amphipods (*Vibilia, Phrosina, Phronima, Phroni-mella, Oxycephalus*), copepods (*Candacia*), decapods (*Sergestes*), euphau-siids (*Meganyctiphanes, Stylocheiron, Thysanoessa*) (Théodoridès and Des-portes, 1975; Kulka and Coney, 1984; Hochberg *in litt.*, 1988).

(ii) *Cephalolobidae. Callyntrochlamys phronimae*, the only known species of this genus, was described by Frenzel (1885) and Dogiel (1910b) and its ultrastructure was studied later by Desportes and Théodoridès (1969). It is an exclusive parasite of hyperiids (*Phronima, Phronimella, Cystisoma*) and has recently been found in the last genus (Hochberg, personal communica-tion). *Cephalolobus lavali* is found in *Phronima* (Théodoridès and Desportes, 1975) but the other species of this genus parasitize benthic decapods.

(iii) *Ganymedidae.* Théodoridès and Desportes (1972) have shown that this family, represented by the genus *Ganymedes* as described originally by

Huxley (1910), includes many species harboured by hyperiids (*Vibilia*) and copepods (*Sapphirina, Calanus, Clausocalanus, Copilia*). These were mentioned earlier as gregarines *incertae sedis* or placed in inadequate genera by various authors (Haeckel, 1864; Mingazzini, 1893; Apstein, 1911; Rose, 1933a; Jepps, 1937; Gobillard, 1963, 1964; see also Soyer (formerly Gobillard), 1965 and Figure 7(A)). In a further paper, Théodoridès and Desportes (1975) mentioned other species of *Ganymedes* from a mysid (*Eucopia*) and a decapod (*Sergestes*). They also described the genus *Lateroprotomeritus*, of unknown taxonomic position, in the euphausiid *Nematoscelis megalops*.

(d) Gregarines from Mollusca

Gregarines have been recorded from marine planktonic prosobranchs (Mesogastropoda) of the family Pterotracheidae (genera *Pterotrachea* and *Firoloida*) (Stuart, 1871; Frenzel, 1885; Trégouboff, 1918; Hochberg and

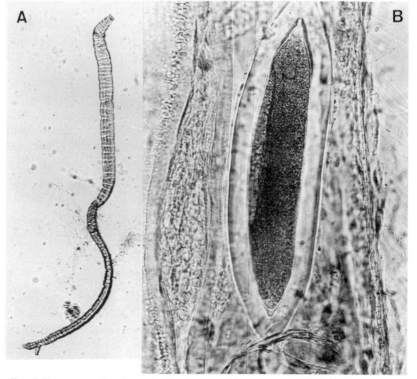

FIG. 7. Two eugregarines from planktonic hosts. (A) *Ganymedes vibiliae* from the intestine of *Vibilia* sp. An association of three trophozoites measuring about 500 μm. (B) Unidentified cyst from *Sagitta punctata* (body cavity) measuring about 340 μm; both hosts are from the plankton of Villefranche-sur-mer (original).

Seapy, 1988). The parasites occur in the digestive tract (oesophagus) and in the body cavity, foot, buccal mass and body wall.

The gregarine found in *Pterotrachea* spp. is *Cephaloidophora clausii* a parasite of copepods (cf. Théodoridès and Desportes, 1975) for which the molluscs seem to be the accidental hosts. This should explain the occurrence of encapsulated trophozoites in either their body cavity or various organs. It was originally described as *Gregarina clausii* (Frenzel, 1885).

In *Firoloida desmaresti*, Hochberg and Seapy (1988) have described *Cephaloidophora firoloidae*.

(e) Gregarines from Tunicata

These parasites occur in the class Thaliacea (Pyrosomidea, Salpidea) and in Appendicularia. They are always found in the intestine.

(i) *Parasites of Pyrosomidea*. Trégouboff (*in* Trégouboff and Rose, 1957) mentions the occurrence of undetermined gregarines in pyrosomes.

(ii) *Parasites of Salpidea*. The gregarines of salps have been known for over a century (Leuckart, 1860; Frenzel, 1885; Roboz, 1886; Bargoni, 1894) and were wrongly placed in the genus *Gregarina*. Ormières (1965) created the genus *Thalicola* for them and later Théodoridès and Desportes (1968) established the new family Thalicolidae with this single genus of which Ormières (1965) and Théodoridès and Desportes (1975) reported three species from the salps off Villefranche-sur-Mer. Among these was *Thalicola ensiformis* which had not been seen since its description by Bargoni (1894).

Some biological features of these gregarines seem to be worth mentioning. They occur in the bent intestine of the hosts and are contained in a protuberance called a nucleus. *Thalicola filiformis*, which has a more elongated shape than the other species of the genus, occurs in *Cyclosalpa virgula* of which the straight intestine is not contained in a nucleus. This seems to be an example of a parasitic adaptation to the host.

In addition to the intestinal trophozoites, Ormières (1965) observed in the salps smooth walled cysts or cysts with clamps localized in the intestinal nucleus, but outside the intestine. These cysts contain trophozoites similar to those present in the intestine.

(iii) *Parasites of Appendicularia*. Fenaux (1963) was the first to report an intestinal gregarine in *Oikopleura albicans*. This is a very small species of which the trophozoites do not exceed 50 μm in length and which has cysts resembling those from *Thalicola* to which it must belong.

2. Coccidia

Coccidia belonging to the genus *Aggregata* (heteroxenous parasites with the schizogony in a decapod crustacean and the gamogony in a cephalopod mollusc) have been found in planktonic decapods (*Gennadas, Sergestes*) from Villefranche-sur-Mer (Théodoridès and Desportes, 1975). The cysts occur in the body cavity on the external surface of the intestine.

I. *Microsporidia*

Although considered by Corliss (1894) and Issi (1986) as a phylum of parasitic protista, the Microsporidia are placed by others outside the latter and near the yeasts.

The only true microsporidian found in a planktonic host seems to be *Plistophora* sp. reported by Chatton (1920) from *Paracalanus parvus*. Dr R. Larsson (personal communication, 1986) is of the opinion that it is very difficult to give a generic name to this parasite. He is also in favour of leaving the Microsporidia in the Protista as they differ from the yeasts in many ways. Although the name *Ichthyosporidium* has been uncritically used to designate a parasitic fungus and a microsporidian, Sprague (1965) suggested that the second designation was the correct one, the fungus being transferred to the genus *Ichthyophonus* (cf. also Sprague, 1966; Sprague and Hussey, 1980; Sprague and Vernick, 1974). As we have seen earlier, this parasite has been found in marine planktonic copepods by Apstein (1911), Chatton (1920) and Jepps (1937) and is therefore mentioned in the present monograph in the Mastigomycetes. An undetermined microsporidian has been reported by Kulka and Corey (1984) as pathogenic for the euphausiid *Thysanoessa inermis* in Canada.

J. *Myxosporidia*

According to Corliss (1984) this assemblage is composed of four groups: Myxosporidia *sensu stricto*, Actinomyxidea, Marteiliidea and Paramyxidea of which only the second and the fourth are present in zooplanktonic marine hosts.

1. Actinomyxidea

Actinomyxidea (or Actinomyxidia) are Cnidosporidia with rather complicated spores composed of three cells. These cells are valved and constitute the

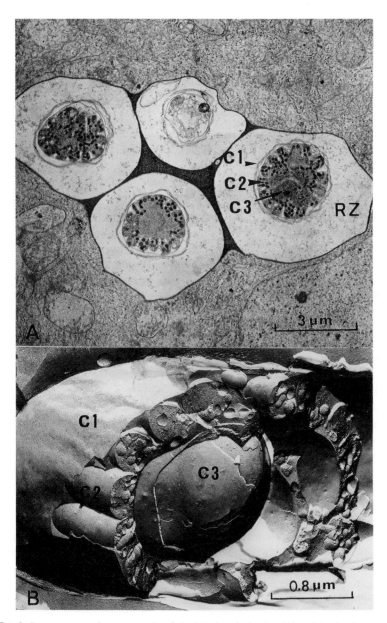

FIG. 8. *Paramyxa paradoxa*, a parasite of planktonic polychaetes. (A) an intratissular sporont containing four mature spores (transverse sections); each spore consists of several cells (C1, C2, C3) (RZ = retraction zone between the outer spore cell (C1) and the sporont wall). (B) Freeze-fracture section of a spore showing the different cells (C1, C2, C3) enclosed inside each other. [(A) After Desportes, unpublished. (B) After Desportes, 1984.]

epispore (or exospore) to which may be joined a unicellular endospore and three polar capsules with filaments. Each valve can be lengthened by long and spiny processes, sometimes bifurcated, which improve buoyancy.

The hosts of these intestinal or coelomic parasites live in either marine (sipunculids), fresh or brackish water (oligochaetes).

At the end of the last century (1894–99), Kofoid (1908) had already noticed the occurrence of spores of Actinomyxidea in the freshwater plankton. His observations were recently confirmed by Marquès and Ormières (1981, 1982) and Marquès (1984) who have shown the adaptation to planktonic life of several species belonging to the genera *Echinactinomyxon*, *Synactinomyxon* and *Aurantiactinomyxon*. The latter three genera are parasitic in aquatic oligochaetes such as *Tubifex*.

Although these observations concern fresh or brackish water hosts, it is almost certain that this adaptation to planktonic life also occurs in parasites of marine hosts (sipunculids) although the latter are benthic.

This is the same biological phenomenon as with metacercariae and cercariae of trematodes found in the marine zooplankton (*vide infra* under the heading Platyhelminthes) i.e. the occurrence in this environment of free-living stages of parasitic organisms.

2. Paramyxidea

This order (also called Paramyxea) is represented by a single genus and a single species (*Paramyxa paradoxa*) parasitic in larvae of planktonic polychaetes (Chatton, 1911b). This intestinal parasite was recently re-discovered in the larvae of *Poecilochaetus serpens* by Desportes and Lom (1981) and Desportes (1981, 1984). They have studied its ultrastructure and showed its affinities with the Marteiliidea parasites of bivalves and amphipods.

Its life cycle consists of the development inside a primary stem cell of many secondary cells (sporonts) giving rise, by endogenous budding, to four infective spores. Each spore is composed of four cells of different sizes enclosed inside each other (Fig. 8).

The effect on the host seems negligible, the parasites being enclosed in a parasitic vacuole and therefore not in direct contact with the host's cells. However the high number of parasites in the gut wall may induce a disorganization of the host's tissue.

V. Animalia

Under this general heading will be considered the metazoan organisms found as parasites in marine zooplanktonic hosts.

A. Hirudinea

In the order Rhynchobdelliformes, family Piscicolidae, the genus *Mysidobdella* is parasitic on mysids. *M. oculata* was described by Selensky (1927) from *Mysis oculata* in the White Sea and found later in other Russian localities and in Greenland by various authors cited by Burreson and Allen (1978).

The latter considered it synonymous with another species of leech (*Ichthyobdella borealis*) and therefore named it *Mysidobdella borealis*. They recorded this parasite from mysids (*Neomysis, Mysis*) of the western North Atlantic and showed that it had a specific preference for *Neomysis americana*.

B. Helmintha

Helminths (parasitic worms), belonging to the Platyhelminthes (Turbellaria, Trematoda, Cestoda) and Nemathelminthes, are frequently found in marine zooplanktonic hosts.

1. Turbellaria

In the order Eulecithophores, family Fecampiidae, a new species has probably been found on *Mysis mixta* in the Gulf of St Lawrence (Canada) (Prof. P. Brunel: personal communication, 1986). Menon (1930) has reported undetermined Turbellaria associated with Scyphomedusae from the Madras coast.

2. Trematoda

Many parasitic flatworms of the subclass Digenea are harboured at the metacercaria stage by planktonic invertebrate hosts or found free-living in the plankton, the definitive hosts being marine fishes (Rebecq, 1965; Dollfus, 1966a,b; Berger *et al.*, 1971; Reimer *et al.*, 1975). In the superfamily Hemiuroidea, these metacercariae may be free or encysted in the intermediate hosts (Lebour, 1923, 1935). Among the latter Dollfus (1923a) has listed ctenophores (*Beroë*), chaetognaths (*Sagitta*) but chiefly copepods (*Acartia, Centropages, Paracalanus, Pseudocalanus, Temora*) in which metacercariae have been found in the body cavity. Other helminth parasites (Trematoda and Nematoda) of ctenophores have been reported from New Zealand (Boyle, 1966) and Ireland (Yip, 1984).

Additional observations have been made by Steuer (1928) and Sewell (1951), the latter in planktonic copepods (*Clausocalanus, Corycaeus*) of the Arabian Sea where metacercariae of *Hemiurus* spp. were observed. Lebour

(1917a) and Dawes (1958) reviewed the larval trematodes reported from *Sagitta* spp. and Dollfus (1960a, 1963) examined those harboured by chaetognaths, cnidarians and ctenarians. More recently Shimazu (1982) reported metacercariae of *Lecithocladium*, *Parahemiurus* and an unnamed hemiurid in the body cavity of Japanese chaetognaths.

Other families of Digenea (Fellodistomidae, Opecoelidae, Accacoelidae, Syncoeliidae, Didymozoidae) also have metacercariae infesting planktonic invertebrates. Examples of this form of parasitism have been given by Dollfus *et al.* (1954), Dollfus (1960a, b), Komaki (1970), Shimazu (1978, 1982) and Nagasawa and Marumo (1979) in relation to chaetognaths and euphausiids. In the latter the Japanese authors report the occurrence of metacercariae of *Tetrochetus*, *Guschanskiana* (Accacoeliidae), *Monilicaecum*, *Torticaecum* (Didymozoidae), *Tergestia* (Fellodistomidae), *Pseudopecoelus* (Opecoelidae) and *Paronatrema* (Syncoeliidae).

Reimer *et al.* (1971) have found four different metacercariae (*Opechona*, *Derogenes*, opecoelid, didymozoid) in various planktonic invertebrates (cnidarians, ctenarians, crustaceans, chaetognaths) from the North Sea.

Weinstein (1972) studied the trematodes *Derogenes varicus* and *Hemiurus levinseni* parasitizing *Sagitta elegans* in the Gulf of St Lawrence (Canada) and showed that they have no effect on the host. As regards the second parasite, he demonstrated that there are very few individuals in the smaller hosts and a constant number of them are found in the 21–40 mm length range of *Sagitta* although they are absent in the 41–50 mm range.

Pearre (1976) suggested that the chaetognaths would be infested while eating copepods (*Paracalanus*, *Acartia*) containing metacercariae. This hypothesis was already made by Apstein (1911) and discussed by Dollfus (1960a). Pearre noticed also that chaetognaths (*Sagitta* spp.) parasitized by *Hemiurus* metacercariae, were longer than those without parasites, the increase varying with the number of parasites per individual host. This was accompanied by a retardation of maturity of the host's ovaries resulting from a partial parasitic castration combined with enhanced somatic growth. S. Dallot (personal communication), however, has not observed any elongated forms in the parasitized *Sagitta* spp. he studied in Villefranche-sur-Mer.

Køie (1979) studied the life cycle of *D. varicus*, infesting calanid copepods experimentally with cercariae from naturally infected molluscs (*Natica* spp.). In *Calanus finmarchicus* they became metacercariae in fishes (plaice and dabs) which were successfully infested. The same author has also found these metacercariae in *S. elegans* and Fig. 9 shows the life cycle of this trematode, involving planktonic invertebrates.

Pearre (1979) reported a niche modification in chaetognaths harbouring larval Digenea, in the sense that the parasitized specimens live nearer the sea surface than those without parasites. Other trematodes of *Sagitta* spp. are

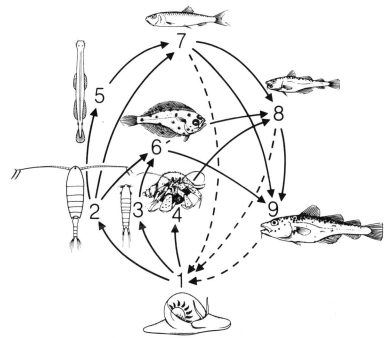

FIG. 9. The life cycle of the trematode *Derogenes varicus* involving planktonic invertebrates. (1) *Natica* spp. (benthic). (2) Calanid copepod. (3) Harpacticoid copepod. (4) Hermit crab (benthic). (5) *Sagitta* spp. (6) Small fishes. (7) Planktophagous fishes. (8) Benthophagous and piscivorous fishes. (9) Piscivorous fishes (large cod). (After Køie, 1979.)

described by Hutton (1954), Furnestin and Rebecq (1966) and Kulachkova (1972).

Yip (1984) has recorded three larval trematodes from *Pleurobrachia pileus* (Ctenophora) in Galway Bay, Ireland. The most abundant was *Opechona bacillaris* (Lepocreadiidae) of which as many as one hundred of metacercariae were found attached to the external wall of the pharynx of a 7 mm host specimen.

The other trematodes found in this host were *Hemiurus communis* (Hemiuridae) and undetermined larvae of Didymozoidae. These parasites do not seem to be very harmful to the host. These observations on *O. bacillaris* confirm those of Køie (1975) who has described the life cycle of this parasite of the mackerel (*Scomber scombrus*). She showed experimentally that its cercariae develop into metacercariae in various planktonic invertebrates (ctenophores, medusae, chaetognaths, etc.) where they have also been found naturally. On the Madras coast, Reimer (1976) found metacercariae of fellodistomatids and lepocreadids in *Pleurobrachia* and leptomedusae. Jarling and Kapp (1985) have found *Cercaria owreae* and *Ectenurus lepidus* in Atlantic chaetognaths (*Sagitta* spp.), whereas Øresland (1986) has recorded *Derogenes, Lecithochirium* and didymozoid metacercariae from *Sagitta setosa* in the western English Channel.

Finally, we must mention the occurrence of metacercariae of *Lepocreadium album* (Lepocreadiidae), a parasite of marine fishes, in the planktonic polychaete *Alciope* sp. (Fuhrmann, 1928; Margolis, 1971), those of another lepocreadiid (*Neonotoporus*) having been found in *Euphausia* in Japan (Shimazu, 1981).

3. Cestoda

The frequent occurrence of larval stages of cestodes (mainly Tetrarhynchidea and Tetraphyllidea), found free in the marine zooplankton or in planktonic invertebrates, has been studied in great detail in an important series of papers by Dollfus (1923a–1976). The larvae found free-living in the plankton belong to the Tetraphyllidea (phyllobothrians) and consist of postlarvae and chains of immature proglottids or are larvae and postlarvae of tetrarhynchids. In all cases the definitive hosts are selachians.

Dollfus (1966a) explained this occurrence in summarizing the life cycle of these parasites which includes a free-living egg and a coracidium larva ingested by a copepod (first intermediate host) in which it becomes a procercoid larva. When the latter is ingested by a second intermediate host (fish, mollusc, crustacean), it becomes a plerocercoid larva which transforms itself into an adult tapeworm in the third and definitive host (a selachian). The postlarvae, found free-living in the plankton, have escaped from the second host and are sometimes able to proceed with their development outside the definitive host (Grabda, 1968). Such is the case of two species described by Dollfus (1966a, 1967) and belonging to the genera *Pseudonybelinia* and *Paranybelinia* from the zooplankton of Boa Vista (Cape Verde Islands) and of the postlarvae of *Oncomegas* sp. found in the plankton of Nhatrang (Vietnam) (Dollfus, 1974).

Other cestode larvae have been found in the body of various planktonic invertebrates ranging from Cnidaria (medusae), Ctenophora (*Beroë, Cydippe, Pleurobrachia*) to Crustacea. Among the latter, *Calanus finmarchicus* harbours a larval tetraphyllid reported by Apstein (1911), found again by Dollfus (1923a) in *C. helgolandicus* but not seen by Jepps (1937) in *C. finmarchicus*. The infestation may be very severe (2600 larvae in a single host). Wundsch (1912) considered these parasites as belonging to two different species of *Plerocercoides*. A tetraphyllid larva has also been found in *Eucalanus* from the Bay of Bengal by Rao and Madhavi (1966).

In euphausiids (*Euphausia, Thysanopoda*), and in the pteropod *Euclio pyramidata*, tetrarhynchids were found among the zooplankton of the tropical western Pacific by Slankis and Shevchenko (1974), whereas in the north Pacific off Japan, Shimazu (1975a, b) reported several larvae and postlarvae of *Nybelinia, Pelichnobothrium, Anomotaenia* in various euphausiids (*Euphausia, Thysanoessa*).

In European and American chaetognaths (*Sagitta* spp.), pseudophyllid larvae have been reported by various authors cited by Dollfus (1974). From the same hosts tetraphyllid plerocercoids were observed as well as larvae of *Echinobothrium* (Diphyllidea) and *Pseudonybelinia* off Japan (Shimazu, 1978, 1982). In the north-western African Atlantic, Reimer (1977) recorded larvae of *Phyllobothrium* spp. and *Scolex pleuronectis* in *Sagitta* spp., *Pterosagitta, Spadella* and *Krohnitta*.

Walter *et al.* (1979) have examined about 18 000 samples of planktonic invertebrates of the White Sea (USSR) and found pseudophyllid procercoids in various copepods (*Acartia, Temora, Pseudocalanus, Centropages*) and in *Sagitta*.

4. Nemathelminthes

It is mainly in planktonic crustaceans (euphausiids, mysids, decapods) that larval nematodes have been found of which the definitive hosts are fishes (Banning, 1967). Such is the case of *Anisakis* sp. and *Contracaecum* sp. observed by Smith (1971) in *Thysanoessa* spp. from the North Sea, *Anisakis* sp. found in Japanese euphausiids (Oshima *et al.*, 1969; Shimazu and Oshima, 1972; Shimazu, 1982) and *A. simplex* observed in the mysid *Mesodopsis slabberi* in Scotland by Makings (1981) and *Contracaecum* sp. in *Erythrops, Mysis* and *Neomysis* in Canada (Scott, 1957).

Slankis and Shevchenko (1974) mention a larval *Philometra* from *Sergestes lucens* and other decapods and euphausiids from the western part of the tropical Pacific, while Lindley (1977) reports larvae of *Anisakis* and *Ascarophis* from the euphausiid *Thysanoessa inermis* in the North Atlantic, thus confirming the above-mentioned observation of Smith (1971).

Walter *et al.* (1979) found third stage larvae of *Contracaecum aduncum* in *S. elegans* from the White Sea and experimentally infested planktonic copepods, *Temora, Acartia* and *Pseudocalanus*, obtaining an infestation of 16.6%.

Shimazu (1982) reported *Thynnascaris* spp. larvae from *Sagitta* spp. and *Calanus sinicus* in Japan.

5. Acanthocephala

The Acanthocephala are intestinal worms having vertebrates as definitive hosts and several intermediate hosts some of which are marine invertebrates, the definitive hosts being fishes, marine birds or mammals.

With regard to zooplanktonic hosts, they have been reported from euphausiids, the first record being that of *Echinorhynchus corrugatus* found in *Euphausia krohni* (Sars, 1885). More recently Shimazu (1975b) noted the occurrence of larvae of *Bolbosoma caenoforme* in the thoracic organs of *Thysanoessa longipes* and *T. raschi* in the northern Pacific off Japan, whereas Lindley (1977) observed larval *Palaeoacanthocephala* in *T. longicauda* from the North East Atlantic.

C. *Pycnogonida*

The Pycnogonids are marine chelicerates whose larvae may develop in cnidarians (hybroids and medusae). Hedgpeth (1962) suggested that bathypelagic species of the genus *Pallenopsis* could be associated with medusae. This hypothesis has been recently confirmed by Allan Child and Harbison (1986) who reported juvenile specimens of *Pallenopsis scoparia* attached to the subumbrellar surface of the scyphomedusa *Periphylla* sp. These authors give evidence that the pycnogonids were feeding on the tentacles of the medusa.

D. *Crustacea*

Many epibiontic or parasitic (ecto- and endoparasites) Crustacea belonging to the three orders Copepoda, Isopoda, Amphipoda occur in the marine plankton.

1. Copepoda

Larval stages of parasitic copepods may be found in larvae or adults of planktonic invertebrates such as annelids, molluscs or crustaceans, playing the role of intermediate hosts. As an example we may mention an unnamed copepod related to the Lernaeidae, of which various stages (copepodites, chalimus and the young sexual forms) are found in planktonic pteropod molluscs (*Criseis, Cavolinia, Clio*) in the plankton of the bays of Algiers and Villefranche-sur-Mer, the definitive host being most probably a fish (Rose and Hamon, 1952).

Other copepods are associates or parasites of invertebrates constituting their definitive hosts and belonging to the Scyphozoa, Mollusca and Crustacea.

(a) *Copepods from Scyphozoa*

Three species of *Paramacrochiron* (Lichomolgidae) have been found associated with medusae (*Lychnorhiza, Rhizostoma*) from India (Reddiah, 1968). The same author (Reddiah, 1969) described a new genus, *Pseudomacrochiron*, associated with another medusa (*Dactylometra*). Some shrimps have also been reported as associated with medusae (Hayashi and Miyake, 1968; Bruce, 1972). The association of cyclopoid copepods with medusae was reported by Humes (1969, 1970) who described a new genus, *Sewellodhiron* (named after Colonel Seymour Sewell, a specialist in Indian Ocean copepods) and a new species of *Paramicrochiron* from medusae of Puerto Rico and Japan. Rose (1933b) described a caligid parasitic in Siphonophora.

(b) *Copepods from Mollusca*

Parasitic copepods have been found in planktonic molluscs, for instance in the three genera *Micrallecto, Nannallecto* and *Pteroxena* described by Stock (1971, 1973), and Stock and Van der Spoel (1976) from pteropods. The first two belong to the Splanchnotrophidae, which is a family that parasitizes molluscs and their hosts (*Pneumoderma, Pneumodermopsis*). The copepods were collected off Bermuda and French Guyana. They are ectoparasites and therefore their morphology is not much modified; however the third genus, which is endoparasitic in *Notobranchaea* has no structured appendages, making it difficult to relate to any known family of copepods.

(c) *Copepods from Crustacea*

Two genera of Choniostomatidae (*Mysidion, Aspidoecia*) are ectoparasitic on the mysids *Erythrops* and *Parerythrops* (Hansen, 1897; Mauchline, 1969, 1980). The parasites, belonging to the first genus (*Mysidion abyssorum, Mysidion commune*), occur in the marsupium of the hosts, whereas *Aspidoecia* is fixed on the abdominal segments, carapace, eyes or in the marsupium.

The free-living harpacticoid *Miracia efferata* has been found associated with the radiolarian *Collozoum longiforme* (Swanberg and Harbison, 1980) whereas copepods (nicothoid, *Hansenulus*) from North American mysids have been reported by Hedon and Damkaer (1986) and Daly and Damkaer (1986).

2. Isopoda

The Epicaridea, a group of parasitic isopods, are common on mysids and euphausiids (Mauchline, 1980). Two families have been reported in mysids: Asconiscidae and Dajiidae, the parasites usually being localized in the marsupium, on the carapace or on abdomen of the host. Among the Asconiscidae only one genus (*Asconiscus*) is a parasite of *Boreomysis*

(Tattersall and Tattersall, 1951). On the other hand, there are several genera of Dajiidae: *Arthrophryxus, Aspidophryxus, Dajus, Holophryxus, Notophryxus, Prodajus* and *Streptodajus* parasitic on various mysids belonging to the genera *Eucopia, Siriella, Parapseudomma, Pseudomma, Amblyops, Gastrosaccus* and *Boreomysis* (Gilson, 1909; Tattersall and Tattersall, 1951; Nouvel, 1951; Taberly, 1954; Hoenigman, 1958, 1960, 1963; Pillai, 1963; Mauchline, 1968, 1970, 1971; Geiger, 1969; Balasubrahmanyan and Prince, 1976).

In euphausiids only members of the Dajiidae, belonging to the genera *Notophryxus, Branchiophryxus* and *Heterophryxus*, are reported as parasites. *Notophryxus* comprises three species: *N. globularis*, parasitic on *Thysanoessa* (North Pacific), *N. lateralis* on *Nematoscelis* (*Nematoscelis megalops* in the South Atlantic and South Pacific and *Nematoscelis difficilis* in the North Pacific) and *Notophryxus lateralis* has been investigated in California by Field (1969) on *Nematoscelis difficilis*. She described the two sexes of the parasite and its larval stages (epicaridian and cryptoniscian larvae). The effect on the host where parasitic castration occurs in both sexes was also studied; males become unable to produce spermatophores or spermatotheca whereas the females have reduced ovaries and undeveloped ovocytes.

Branchiophryxus comprises three species: *B. nyctiphanae* which is a parasite of *Meganyctiphanes norvegica* discovered by Caullery (1897) and seen again by Masi (1905), *B. koehleri* is parasitic on *Stylocheiron carinatum* (Sebastian, 1970) and *B. caulleryi* on *S. longicorne* (Koehler, 1911).

Heterophryxus appendiculatus is a parasite of *Euphausia krohni* and *E. sibogae* (Sars, 1885; Sebastian, 1970; Lindley, 1977) of which a detailed description has been given by Drago and Albertelli (1975).

3. Amphipoda

Among amphipods two main groups, mostly the gelatinous hyperiids and caprellids, are found in association with planktonic invertebrates.

(a) *Hyperiids*

All the members of this suborder are associated with gelatinous zooplankton (Radiolaria, Cnidaria, Siphonophora, Ctenophora, Mollusca, Tunicata) in their early stages. They are defined as "parasitoids" by Laval (1980) who has described the life cycle of several genera such as *Vibilia, Hyperia, Phronima* and *Bougisia*. Generally the eggs hatch in the maternal marsupium (egg-pouch) and the first larvae are transferred from there to the host (demarsupiation) by the female. In *Bougisia* and *Hyperia* the eggs are laid directly into the leptomedusa host (P. Laval, 1966, 1972).

Once they develop into adults, the hyperiids leave their host and become free living. However, in the case of *Phronima*, the female and its offspring remain in a barrel-shaped case made by the tissue from the salp host (Laval, 1978) (Fig. 10 (B)).

The main families of hyperiids and their respective hosts are: Lanceolidae (medusae), Scinidae and Vibiliidae (tunicates), Paraphronimidae (siphonophores), Hyperiidae (radiolarians, medusae and ctenophores), Phrosinidae (tunicates), Phronimidae (siphonophores and tunicates), Lycaeopsidae (siphonophores), Pronoidae (medusae and siphonophores), Lycaeidae (medusae, tunicates, molluscs), Oxycephalidae (ctenophores), Platyscelidae (medusae and siphonophores) and Parascelidae (siphonophores) (Laval, 1980).

The main recent works on the biology of hyperiids are by Flores and Brusca (1975), Harbison (1976), Thiel (1976), Harbison *et al.* (1977, 1978), Madin and Harbison (1977) and Swanberg and Harbison (1980) for the Atlantic species and P. Laval (1963, 1965, 1968a, b, 1974a,b, 1975, 1978, 1980) for the Mediterranean species. An association between *Hyperia galba* and *Desmonema gaudichaudi* was studied by White and Bone (1972). The life cycle of two Atlantic species of *Lycaea* has been described, as well as the

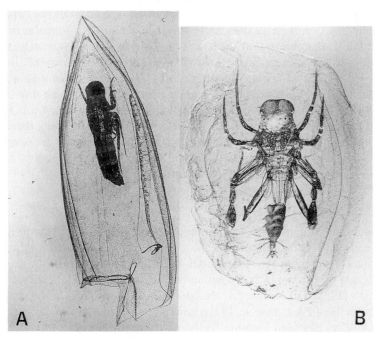

A   B

FIG. 10. Association of hyperiid crustaceans with gelatinous zooplankton. (A) Subadult male of *Lycaeopsis themistoides* in a nectophore of *Chelophyes appendiculata* (siphonophore) (size: 13.6 mm). (B) Subadult female of *Phronima atlantica* in its barrel shaped from an oozoid of *Salpa fusiformis* (size: 15.5 mm). (After P. Laval, 1974b.)

association of various hyperiids with cnidarians, ctenophores and salps (Harbison *et al.*, 1977).

A peculiar choice of host is shown by the occurrence of various hyperiids (*Hyperietta* spp., *Oxycephalus*, *Brachyscelus*, *Lycaea*) in the gelatinous colonies of a radiolarian (*Collozoum longiforme*) which can attain large dimensions (Swanberg and Harbison, 1980). In the Mediterranean species, Laval (1963) showed that the larvae of *Vibilia* were deposited by the female on various tunicates (*Thalia*, *Salpa*, *Ihlea*), on which they later developed. A special device is observed in these larvae which have styliform prolongations on their pereiopods facilitating their attachment to the host. The life cycle and development of the various hyperiid genera, *Vibilia*, *Hyperia*, *Bougisia* and *Lycaeopsis*, have been investigated by Laval (1965) who further discovered, described and studied the new genus *Bougisia* whose early stages are parasitic in a leptomedusa, *Phialidium* (Laval, 1966).

The same author (P. Laval, 1968a) described the biology of *Phronima curvipes* where the female lives in a barrel-shaped case comprising a nectophore of the siphonophore *Abylopsis tetragona*. He also studied the life cycle of *Hyperia* (*Lestrigonus*) *schizogeneios* where the larvae were deposited by the female in the host, *Phialidium* sp. (Laval, 1968b). The effect of the hyperiids on their hosts, at least in their early stages, is definitely a harmful one. P. Laval (1972) has shown that the larvae of *Hyperia schizogeneios* will eat the gonads and manubrium of the host (*Phialidium*) and therefore kill it. The same author (Laval, 1974b) reported that *Lycaeopsis themistoides* ate the ectoderm of the siphonophore *Chelophyes appendiculata* damaging it seriously (Fig. 10 (A)). The transformation of the oozoids of *Salpa* spp. into a modified barrel by the female of *Phronima* spp. and the breeding therein of their offspring (as many as 600 larvae in the case of *Phronima sedentaria*) is one of the most astonishing ethological features of the hyperiids and is unique among zooplanktonic invertebrates (Laval, 1974b).

(b) *Other amphipods*

In his review of gammarid and caprellid amphipods associated with medusae, Vader (1972b) mentioned the occurrence of *Metopa alderii* (Stenothoidae) on *Timea bairdii*. The amphipods are found on all parts of the host but most often on the stomach and subumbrella. The same author reported that Elmhirst (1925) had observed a related species (*Metopa borealis*) on or under the umbrella of *Phialidium* sp. In the same way *Panopea eblanae* (Acanthonotozomatidae) has been found in the branchial cavities of *Rhizostoma pulmo*. Among the caprellids, Hamond (1967) recorded the occurrence of *Pariambus typicus* on *Rhizostoma* sp. The relation of these amphipods to their hosts is quite different from those of the hyperiids and can be defined as inquilinism i.e. they use them for food and shelter without damaging them in any way.

E. *Mollusca*

Two species of gastropod molluscs have been reported as ectoparasitic on medusae and siphonophores. They belong to the genera *Phylliroe* and *Cephalopyge* (Opisthobranchia, Dendronotacea, Phylliroidae). *Phylliroe bucephalum* is parasitic on a medusa (*Zanclea costata*) which had earlier been wrongly considered as a parasite of the mollusc. Its life cycle has been described by Martin and Brinckmann (1963) who showed that, at the veligerous stage, *Phylliroe* penetrated into the umbrella of a young medusa where it grew and then protruded outside, having eaten its tentacles and umbrella.

Cephalopyge trematoides has been reported in Villefranche-sur-Mer as ectoparasites on the siphonophore, *Nanomia bijuga*, to which the young molluscs are attached by their foot glands, eating the hosts' tentillae, zooids and stolon (Sentz-Braconnot and Carré, 1966).

VI. Conclusion

In his monograph on the parasites of *Sagitta elegans*, Weinstein (1972) rightly remarked:

> "Planktonic organisms, like other organisms, provide an environment, complete with special advantages and inherent disadvantages for potential colonizers. The dearth of available information on the parasites of marine zooplanktonic species does not necessarily reflect absence or low diversity of parasites in these animals."

The same author further notes:

> "There is a large literature on parasitized zooplankters, but because of the huge number of possible host-parasite taxa combinations, a complete review would be a monumental task."

In the same work, Weinstein recalls that, until 1972, only reviews limited to one zooplankton or parasite taxon had been published and he quotes examples from Dollfus's papers (1923–1967) on helminths from planktonic hosts or the monographs of Jepps (1937) and Sewell (1951) on the parasites of planktonic copepods.

The situation has hardly changed since 1972 and this is why I have tried to give a rather complete, if not exhaustive, account of the main groups of microorganisms and organisms reported from zooplanktonic hosts here. To this end I have quoted the main references available, stressing the papers published since 1950. It will be obvious from the foregoing text that the greatest number of parasites of zooplanktonic hosts are from the Protista and its assemblages. Among these, the parasitic dinoflagellates (peridinians) are particularly numerous, occurring in the various components of the zooplankton from Protista to fishes and showing morphological or biological adaptations in relation to parasitism from epibiontism to endoparasitism (Cachon and Cachon, 1964–1987).

That curious group, the Ellobiopsidae, the taxonomic position of which is still unclear and was first considered as harmless epibionts of planktonic Crustacea, are often endoparasites affecting the fertility of their hosts (Mauchline, 1966).

In the Ciliophora from zooplanktonic hosts, all the life styles from epibiontism to endoparasitism are also observed, while the Sporozoa (Gregarinia, Coccidia) are all endoparasitic. In the Actinomyxidea, special morphological adaptations to planktonic life are found, consisting of spiny processes of spores facilitating their buoyancy in the water (Marquès and Ormières, 1981, 1982).

Among the Metazoan parasites of zooplanktonic hosts, the most important and interesting are the platyhelminths (trematodes, cestodes) of which the definitive hosts are fishes. Their larval stages, cercariae, metacercariae, coracidia, procercoids and plerocercoids, develop in planktonic invertebrates such as Ctenophora, Polychaeta, Chaetognatha and Crustacea, which act as intermediate hosts, or they are found free living in the plankton (Dollfus, 1966a, b). This last localization has somehow modified the previous conceptions on the life cycle and developmental stages of platyhelminths.

As in the previously mentioned groups of parasites, the relations between the parasitic Crustacea and their zooplanktonic hosts range from epibiontism to endoparasitism. Such is the case in the copepods, associated with medusae, molluscs and other crustaceans such as mysids. Among the isopods the epicarids are frequently found on mysids and euphausiids (Mauchline, 1980). However, the most peculiar and interesting factor in the relationship between Crustacea with zooplankton is that the hyperiid amphipods are all associated, in their early stages, with gelatinous groups of hosts, from radiolarians to tunicates. They have been defined as "parasitoids" (Laval, 1980), often having a harmful action on their hosts which they may eventually kill. Another feature of the hyperiids is their ability to build "barrels" from siphonophore nectophores or tunicate (salp) oozoids in which the female lives with its offspring (Laval, 1974b).

There are only a very few studies on the possible harmful role of parasites on their zooplanktonic hosts. Weinstein (1972) examined about 10 000 individuals of *Sagitta elegans* in plankton of the southwestern Gulf of St Lawrence. This chaetognath is parasitized by a ciliate (*Metaphyra sagittae*), two trematodes (*Hemiurus levinsi, Derogenes varicus*) and nematodes (*Contracaecum* spp.). From these observations it does not appear that these parasites have any pathogenic effect on their host and thus they play no role in the fluctuations of its populations. Further studies on other groups of planktonic hosts and their parasites are much needed and may yield different results.

Acknowledgements

I wish to thank here all those who have, in many ways, helped in the preparation of this monograph. First of all, Professors P. Bougis, P. Nival and the staff at the Station zoologique of Villefranche-sur-Mer (Alpes-Maritimes, France) of the Paris Université P. & M. Curie, where I stayed many times from 1966 to 1978.

Specialists in the different parasite groups have also been a great help, providing documentation and critical comments. They are: Dr M. Moreau-Peuto (formerly M. Laval) (Bacteria), Dr F. C. Page (Rhizopoda), Dr G. Brugerolle (Polymastigotes), Professor J. and Mrs M. Cachon, Professor A. Hollande and Dr M. O. Soyer-Gobillard (Dinoflagellata), Professor R. Hovasse and Dr J. B. Crumeyrolle (Ellobiopsidea), Professor J. Corliss and Dr A. Batisse (Ciliophora), Dr F. G. Hochberg (Sporozoa), Dr R. Larsson (Microsporidia), Dr I. Desportes (Paramyxea), the late Professor R. Ph. Dollfus and Dr T. Shimazu (Helminths), Professor Th. Monod, Dr Ph. Laval and Dr J. Stock (Crustacea).

Special thanks are due to Professor P. Brunel (Montreal) for the loan of his copy of the unpublished thesis of Dr M. Weinstein and to the latter who has allowed me to quote from it. Also to Mrs J. Carpine-Lancre for invaluable bibliographical assistance from the Library of the Musée Océanographique (Monaco) and Mr D. Morineau for photographic help. I am also grateful to the editors of *Advances in Marine Biology* who have kindly accepted the manuscript for publication in spite of the long delay in its delivery and especially to Dr J. H. S. Blaxter who has so carefully read and corrected it. Also to an anonymous referee who has suggested several valuable modifications and additions.

References

Allan Child, C. and Harbison, G. R. (1986). A parasitic association between a pycnogonid and a scyphomedusa in midwater. *Journal of the Marine Biological Association of the United Kingdom* **66**, 113–117.
Alldredge, A. and Jones, B. (1973). *Hastigerina pelagica*: foraminiferal habitat for planktonic dinoflagellates. *Marine Biology* **22**, 131–135.
Anderson, O. R. (1976a). Ultrastructure of a colonial radiolarian *Collozoum inerme* and a cytochemical determination of the role of its zooxanthellae. *Tissue and Cell* **8**, 195–208.
Anderson, O. R. (1976b). A cytoplasmic fine-structure study of two spumellarian Radiolaria and their symbionts. *Marine Micropaleontology* **1**, 81–99.
Anderson, O. R. and Bé, A. W. H. (1976). The ultrastructure of a planktonic foraminifer, *Globigerinoides sacculifer* (Brady) and its symbiotic dinoflagellates. *Journal of Foraminiferal Research* **6**, 1–21.
Apstein, C. (1911). Parasiten von *Calanus finmarchicus*. *Wissenschaftliche Meeresuntersuchungen, Neue Folge* **13**, 211–213.

Balasubrahmanyan, K. and Prince, M.J. (1976). *Prodajus ovatus* Pillai (Isopoda) parasitic on a new host. *Current Science* **45**, 603.

Balderston, W. L. and Claus, G. (1969). A study of the symbiotic relationship between *Symbiodinium microadriaticum* F. a zooxanthella and the upside down jellyfish *Cassiopeia* sp. *Nova Hedwigia* **17**, 373–382.

Banning, P. van (1967). Nematodes in plankton samples from the North Sea. *International Council of the Exploration of the Sea* **20**, 1–4.

Bargoni, E. (1894). Di un foraminifero parassita nelle Salpe (*Salpicola amylacea* n.g.n.sp.) e considerazioni sui corpusculi amilacei dei protozoi superiori. *Ricerche fatte nel Laboratorio di Anatomia normale della Reale Università di Roma* **4**, 43–62.

Bauld, J. and Staley, J. T. (1976). *Planctomyces maris* sp. nov.: a marine isolate of the *Planctomyces–Blastocaulis* group of budding bacteria. *Journal of General Microbiology* **97**, 45–55.

Berger, C., Heuer, B. Lainka, H., Rosenthal, I. and Scharnweber, I. (1971). On the distribution of larvae of helminths in plankton animals of the North Sea (in Russian). *Parazitologia* **5**, 542–550.

Bergoin, M., Mialhe, E., Quiot, J. M. and Van Rosieren, P. (1984). Infection à *Chloriridovirus* (Iridoviridae) dans les populations du Crustacé planctonique *Daphnia magna* (Straus) (Cladocera) d'étangs saumâtres. *Comptes-rendus de l'Académie des Sciences, Paris, Série A* **298**, 139–142.

Berkeley, C. (1930). Symbiosis of a *Beroë* and a flagellate. *Contributions to Canadian Biology and Fisheries* **6**, 13–21.

Bhaldraithe, P. de (1973). A record of the ellobiopsid *Thalassomyces fagei* (Boschma 1948), parasitic on the euphausiid *Meganyctiphanes norvegica* (M. Sars 1856). *Bulletin of the Zoological Museum of the University of Amsterdam* **3**, 69–71.

Boschma, H. (1948). Sur les organelles d'absorption chez une espèce d'*Amallocystis* (Protozoa, Ellobiopsidae) *Proceedings Koninklijke Nederlandsche Akademie van Wetenschapen* **51**, 446–449.

Boschma, H. (1949). Ellobiopsidae. *Discovery Reports* **25**, 281–314.

Boyle, M. S. (1966). Trematode and nematode parasites of *Pleurobrachia pileus* O. F. Müller in New Zealand waters. *Transactions of the Royal Society of New Zealand, Zoology* **8**, 52–62.

Brown, E. M. (1931). Note on a species of dinoflagellate from the gills and epidermis of marine fishes. *Proceedings of the Zoological Society of London* **30**, 345–346.

Brown, E. M. (1934). On *Oodinium ocellatum* Brown, a parasitic dinoflagellate causing epidermic disease in marine fishes. *Proceedings of the Zoological Society of London* **33**, 583–607.

Brown, E. M. and Hovasse, R. (1946). *Amyloodinium ocellatum* (Brown) a peridinian parasite on marine fishes. A complementary study. *Proceedings of the Zoological Society of London, N.S.* **116**, 33–46.

Bruce, A. J. (1972). An association between a pontoniinid shrimp and a rhizostomatous scyphozoan. *Crustaceana* **23**, 300–302.

Brugerolle G. and Charnier, M. (1981). Ultrastructure du Bodonidé *Trypanophis grobbeni* Poche. *Protistologica* **8**, 223–236.

Burreson, E. M. and Allen, D. M. (1978). Morphology and biology of *Mysidobdella borealis* (Johansson) comb. nov. (Hirudinea, Piscicolidae) from mysids in the western North Atlantic. *Journal of Parasitology* **64**, 1082–1091).

Cachon, J. (1953). Morphologie et cycle évolutif de *Diplomorpha paradoxa* (Rose et Cachon) Péridinien parasite des Siphonophores. *Bulletin de la Société Zoologique de France* **78**, 408–414.

Cachon, J. (1964). Contribution à l'étude des Péridiniens parasites, cytologie, cycles évolutifs. *Annales des Sciences Naturelles, Zoologie 12e Série* **6**, 1–158.
Cachon, J. and Cachon-Enjumet, M. (1964). Cycle évolutif et cytologie de *Neresheimeria catenata* Neresh. Péridinien parasite d'Appendiculaires. Rapports de l'hôte et du parasite. *Annales des Sciences Naturelles, Zoologie 12e Série* **6**, 779–800.
Cachon, J. and Cachon, M. (1965). *Atlanticellodinium tregouboffi* nov. gen. nov. sp. Péridinien Blastuloidae Neresheimer, parasite de *Planktonetta atlantica* Borgert, Phaeodarié Atlanticellide. Cytologie, cycle évolutif, évolution nucléaire au cours de la sporogenèse. *Archives de Zoologie expérimentale et générale* **105**, 369–379.
Cachon J. and Cachon, M. (1966). Ultrastructure d'un Péridinien parasite d'Appendiculaires, *Neresheimeiria catenata* (Neresheimer). *Protistologica* **2**, 17–25.
Cachon, J. and Cachon, M. (1968). Cytologie et cycle évolutif des *Chytriodinium* (Chatton). *Protistologica* **4**, 249–262.
Cachon, J. and Cachon, M. (1969a). Cycles évolutifs de quelques Péridiniens parasites de la tribu des Duboscquodinida (Chytriodinidae et Blastuloidae). *Progress in Protozoology*, Leningrad, p. 381.
Cachon, J. and Cachon, M. (1969b). Ultrastructures des Amoebophryidae (Péridiniens: Duboscquodinida). I. Manifestations des rapports entre l'hôte et le parasite. *Protistologica* **5**, 535–547.
Cachon, J. and Cachon, M. (1971a). *Protoodinium chattoni* Hovasse. Manifestations ultrastructurales des rapports entre le péridinien et la Méduse-hôte: fixation, phagocytose. *Archiv für Protistenkunde* **113**, 293–305.
Cachon, J. and Cachon, M. (1971b). Ultrastructures du genre *Oodinium* Chatton. Différenciations cellulaires en rapport avec la vie parasitaire. *Protistologica* **7**, 153–169.
Cachon, J. and Cachon, M. (1973). Les Apodinidae Chatton. Révision systématique. Rapports hôte-parasite et métabolisme. *Protistologica* **9**, 17–33.
Cachon, J. and Cachon, M. (1974a). Comparaison de la mitose des Péridiniens libres et parasites à propos de celle des *Oodinium*. *Comptes-rendus de l'Académie des Sciences, Paris* **278**, 1735–1737.
Cachon, J. and Cachon, M. (1974b). Le système stomopharyngien de *Kofoidinium* Pavillard. Comparaisons avec celui de divers Péridiniens libres et parasites. *Protistologica* **10**, 217–222.
Cachon, J. and Cachon, M. (1977). Observations on the mitosis and on the chromosome evolution during the life-cycle of *Oodinium*, a parasitic dinoflagellate. *Chromosoma* **60**, 237–251.
Cachon, J. and Cachon, M. (1987). Parasitic Dinoflagellates. *In* "The Biology of Dinoflagellates" (F. J. R. Taylor ed.), pp. 571–610. Blackwell, Oxford.
Cachon, J., Cachon, M. and Bouquaheux, F. (1965). *Stylodinium gastrophilum* Cachon, Péridinien Dinococcide parasite de Siphonophores. *Bulletin de l'Institut Océanographique* **65**, no. 1359, 1–8.
Cachon, J., Cachon, M. and Pyne, C. (1968). Structure et ultrastructure de *Paradinium poucheti* (Chatton 1910) et position systématique des Paradinides. *Protistologica* **4**, 303–311.
Cachon, J., Cachon, M. and Greuet, C. (1970). Le système pusulaire de quelques Péridiniens libres ou parasites. *Protistologica* **6**, 467–476.
Cachon, J., Cachon, M. and Charnier, M. (1972). Ultrastructure du Bodonidé *Trypanophis grobbeni* Poche parasite de Siphonophores. *Protistologica* **8**, 223–236.
Cachon, M. and Caram, B. (1979). A symbiotic green alga, *Pedinomonas symbiotica*

sp. nov. (Prasinophyceae) in the radiolarian *Thalassolampe margarodes*. *Phycologia* **18**, 177–184.

Canter, H. M. (1959). Fungal parasites of the phytoplankton. IV. *Rhizophydium contractophilum* sp. nov. *Transactions of the British Mycological Society* **46**, 306–320.

Carré, C. (1968). Contribution à l'étude du genre *Sphaeronectes* Huxley 1859. *Vie et Milieu, Série A* **19**, 85–94.

Caullery, M. (1897). *Branchiophryxus nyctiphanae* n.g.n.sp. épicaride nouveau de la famille des Dajidae. *Zoologische Anzeiger* **20**, 88–92.

Caullery, M. (1910). *Ellobiopsis chattoni* n.g.n.sp. parasite de *Calanus helgolandicus* Claus, appartenant probablement aux péridiniens. *Bulletin Scientifique de la France et de la Belgique* **44**, 201–214.

Caullery, M. and Mesnil, F. (1905). Recherches sur les Haplosporidies. *Archives de Zoologie expérimentale et générale* **4**, 101–181.

Cépède, C. (1910). Recherches sur les Infusoires astomes. *Archives de Zoologie expérimentale et générale, 5e série* **3**, 341–609.

Chatton, E. (1906). Les Blastodinides, ordre nouveau de Dinoflagellés parasites. *Comptes-rendus de l'Académie des Sciences* **143**, 981–983.

Chatton, E. (1911a). Ciliés parasites des Cestes et des Pyrosomes: *Perikaryon cesticola* n.g.n.sp. et *Conchophrys davidoffi* n.g.n.sp. *Archives de Zoologie expérimentale et générale, 5e série* **8**, N. & R., VIII–XX.

Chatton, E. (1911b). Sur une Cnidosporidie sans cnidoblaste: *Paramyxa paradoxa* n.g.n.sp. *Comptes-rendus de l'Académie des Sciences, Paris* **152**, 631–633.

Chatton, E. (1912). Diagnoses préliminaires de Péridiniens parasites. *Bulletin de la Société Zoologique de France* **37**, 85–92.

Chatton, E. (1920). Les Péridiniens parasites. Morphologie, reproduction, éthologie. *Archives de Zoologie expérimentale et générale* **59**, 1–475.

Chatton, E. (1927). La gamétogenèse méiotique du Flagellé *Paradinium poucheti*. *Comptes-rendus de l'Académie des Sciences* **185**, 400–403.

Chatton, E. (1940). Le chondriome périnucléaire et radiaire du Rhizopode protéomyxé *Paradinium poucheti*. *Comptes-rendus de la Société de Biologie* **134**, 232–233.

Chatton, E. (1952). Classe des Dinoflagellés ou Péridiniens. *In* "Traité de Zoologie" (P. Grassé ed.), I, fasc. 1, pp. 309–390, Masson, Paris.

Chatton, E. and Biecheler, B. (1934). Les *Coccidinidae*, Dinoflagellés coccidiomorphes parasites de Dinoflagellés et le phylum des *Phytodinozoa*. *Comptes-rendus de l'Académie des Sciences, Paris* **198**, 252–255.

Chatton, E. and Biecheler, B. (1935). Les *Amoebophrya* et le *Hyalosaccus*; leur cycle évolutif. L'ordre nouveau des *Coelomastigina* dans les Flagellés. *Comptes-rendus de l'Académie des Sciences, Paris* **200**, 505–507.

Chatton, E. and Biecheler, B. (1936). Documents nouveaux relatifs aux Coccidinides (Dinoflagellés parasites). La sexualité du *Coccidinium mesnili* n.sp. *Comptes-rendus de l'Académie des Sciences, Paris* **203**, 573–575.

Chatton, E. and Hovasse, R. (1937). *Actinodinium apsteini* n.g.n.sp. Péridinien parasite entérocoelomique des *Acartia* (Copépodes). *Archives de Zoologie expérimentale et générale* **79**, N. & R., 24–29.

Chatton, E. and Lwoff, A. (1935). Les Ciliés Apostomes. I. Aperçu historique et général; étude monographique des genres et des espèces. *Archives de Zoologie expérimentale et générale* **77**, 1–453.

Chatton, E. and Lwoff, A. (1939). Sur la systématique de la tribu des Thigmotriches Rhynchoidés. Les deux familles des *Hypocomidae* Bütschli et des *Ancistrocomidae* n.fam. Les deux genres nouveaux *Heterocoma* et *Parhypocoma*. *Comptes-rendus de l'Académie des Sciences, Paris* **209**, 429–431.

Chatton, E. and Soyer, M. O. (1973). Le cycle évolutif de *Paradinium poucheti* Chatton, Flagellé parasite plasmodial des Copépodes. Les Paradinides. *Annales des Sciences Naturelles Zoologie, 12e série* **15**, 27–60.

Collin, B. (1912). Etude monographique sur les Acinétiens. II. Morphologie, Physiologie, Systématique. *Archives de Zoologie expérimentale et générale* **51**, 1–457.

Corbel, J. C., Desportes, I. and Théodoridès, J. (1979). Etude de *Gonospora beloneides* (Ming.) (= *Lobianchella beloneides* Ming.) parasite coelomique d'une Alciope (Polychaeta) et remarques sur d'autres Grégarines d'Alciopidae. *Protistologica* **15**, 55–65.

Corliss, J. O. (1979). "The Ciliated Protozoa: characterization, classification and guide to the literature". 455 pp. Pergamon Press, Oxford.

Corliss, J. O. (1984). The Kingdom Protista and its 45 phyla. *Biosystems* **17**, 87–126.

Corliss, J. O. (1986). The kingdoms of organisms—From a microscopist's point of view. *Transactions of the American Microscopical Society* **105**, 1–10.

Coutière, H. (1911). Les Ellobiopsidae des crevettes bathypélagiques. *Bulletin Scientifique de France et de Belgique* **45**, 186–206.

Dales, R. P. (1966). Symbiosis in marine organisms. *In* "Symbiosis, 1. Associations of microorganisms, plants and marine organisms" (S. M. Henry, ed.), pp. 299–326. Academic Press, New York & London.

Daly, K. L. and Damkaer, D. M. (1986). Population dynamics and distribution of *Neomysis mercedis* and *Alienacanthomysis macropsis* (Crustacea Mysidacea) in relation to the parasitic copepod *Hansenulus trebax* in the Columbia river estuary. *Journal of Crustacean Biology* **6**, 840–857.

Dawes, B. (1958). *Sagitta* as a host of larval Trematodes, including a new and unique type of cercaria. *Nature* **182**, 960–961.

Dawes, B. (1959). On *Cercaria owreae* (Hutton 1954) from *Sagitta hexaptera* (d'Orbigny) in the Caribbean plankton. *Journal of Helminthology* **33**, 209–222.

Desportes, I. (1981). Etude ultrastructurale de la sporulation de *Paramyxa paradoxa* Chatton (Paramyxida) parasite de l'Annélide Polychète *Poecilochaetus serpens*. *Protistologica* **17**, 365–386.

Desportes, I. (1984). The Paramyxea Levine 1979: an original example of evolution towards multicellularity. *Origins of Life* **13**, 343–352.

Desportes, I. and Lom, J. (1981). Affinités de *Paramyxa paradoxa* Chatton 1911, parasite de *Poecilochaetus serpens* (Annélide Polychète) avec les Marteiliidae Sprague, parasites d'Huîtres et du Crustacé *Orchestia gammarellus*. *Comptes rendus de l'Académie des Sciences* **292**, 627–632.

Desportes, I. and Théodoridès, J. (1969). Ultrastructure de la Grégarine *Callyntrochlamys phronimae* Frenzel; étude comparée de son noyau avec celui de *Thalicola salpae* (Frenzel) (Eugregarina). *Journal of Protozoology* **16**, 449–460.

Dogiel, V. (1910a). Untersuchungen über einige neue Catenata. *Zeitschrift für wissenschaftliche Zoologie* **94**, 400–446.

Dogiel, V. (1910b). Beiträge zur Kenntnis der Gregarinen. IV *Callyntrochlamys phronimae* Frenzel. *Archiv für Protistenkunde* **20**, 60–78.

Dollfus, R.Ph. (1923a). Enumération des Cestodes du plancton et des Invertébrés marins. *Annales de Parasitologie humaine et comparée* **1**, 276–300; 363–394.

Dollfus, R.Ph. (1923b). Remarques sur le cycle évolutif des Hémiurides. *Annales de Parasitologie humaine et comparée* **1**, 345–351.

Dollfus, R.Ph. (1924). Enumération des Cestodes du plancton et des Invertébrés marins. *Annales de Parasitologie humaine et comparée* **2**, 86–89.

Dollfus, R.Ph. (1931). Nouvel addendum à mon "énumération des Cestodes du plancton et des Invertébrés marins". *Annales de Parasitologie humaine et comparée* **9**, 552–560.

Dollfus, R.Ph. (1960a). Distomes des Chaetognathes. *Bulletin de l'Institut des Pêches maritimes du Maroc* **4**, 19–45.

Dollfus, R.Ph. (1960b). Critique de récentes innovations apportées à la classification des Accacoeliidae (Trematoda Digenea). Observations sur des métacercaires de cette famille. *Annales de Parasitologie humaine et comparée* **35**, 648–671.

Dollfus, R.Ph. (1963). Liste des Coelentérés marins paléarctiques et Indiens où ont été trouvés des Trématodes digénétiques. *Bulletin de l'Institut des Pêches maritimes du Maroc* **9–10**, 33–57.

Dollfus, R.Ph. (1964). Enumération des Cestodes du plancton et des Invertébrés marins (6e contribution). *Annales de Parasitologie humaine et comparée* **39**, 329–379.

Dollfus, R.Ph. (1966a). Organismes dont la présence dans le Plancton marin était jusqu'à présent ignorée: larves et postlarves de Cestodes Tétrarhynques. *Comptes-rendus de l'Académie des Sciences, Paris* **262**, 2612–2615.

Dollfus, R.Ph. (1966b). Métacercaire énigmatique de distome du plancton de surface des Iles du Cap Vert. *Bulletin du Muséum national d'Histoire naturelle, 2e série* **38**, 195–200.

Dollfus, R.Ph. (1967). Enumération des Cestodes du plancton et des Invertébrés marins (7e contribution). *Annales de Parasitologie humaine et comparée* **42**, 155–178.

Dollfus, R.Ph. (1974). Enumération des Cestodes du plancton et des Invertébrés marins. 8e contribution avec un appendice sur le genre *Oncomegas* R.Ph.D. 1929. *Annales de Parasitologie humaine et comparée* **49**, 381–410.

Dollfus, R.Ph. (1976). Enumération des Cestodes du plancton et des Invertébrés marins. 9e contribution. *Annales de Parasitologie humaine et comparée* **51**, 207–220.

Dollfus, R.Ph., Anantaraman, M. and Volappan Nair, R. (1954). Métacercaire d'Accacoeliidé chez *Sagitta inflata* Grassi et larve de Tétraphyllide fixée à cette métacercaire. *Annales de Parasitologie humaine et comparée* **29**, 521–526.

Dragesco, J. (1948). Sur la biologie du *Zoothamnium pelagicum* (du Plessis) (Note préliminaire). *Bulletin de la Société Zoologique de France* **73**, 130–134.

Drago, N. and Albertelli, G. (1975). Contributo allo studio di *Heterophryxus appendiculatus* G. O. Sars. *Cahiers de Biologie marine* **16**, 435–443.

Drebes, G. (1969). *Dissodinium pseudocalani* sp.nov. ein parasitischer Dinoflagellat auf Copepodeneiern. *Helgoländer wissenschaftliche Meeresuntersuchungen* **19**, 58–67.

Drebes, G. (1972). *Dissodinium pseudocalani* (Dinophyceae)—Vegetative Vermehrung. *Begleitveröffentlichen zum Film des Instituts des wissenschaftlichen Films*, E 1634, Göttingen, 1–11.

Drebes, G. (1978). *Dissodinium pseudolunula* (Dinophyta) a parasite on copepod eggs. *British Phycological Journal* **13**, 319–327.

Droop, M. R. (1963). Algae and invertebrates in symbiosis. In "Symbiotic Associations" (P. S. Nutman and B. Mosse, eds.), pp. 171–199, Cambridge University Press, Cambridge.

Duboscq, O. and Rose, M. (1926). *Trypanophis major* n.sp. parasite d'*Abylopsis pentagona*. *Bulletin de la Société Zoologique de France* **51**, 372–376.

Duboscq, O. and Rose, M. (1927). Les stades grégariniens et les kystes de *Trypano-*

phis major Duboscq et Rose. *Bulletin de la Société d'Histoire naturelle d'Afrique du Nord* **18**, 94–96.

Duboscq, O. and Rose, M. (1933). *Trypanophis grobbeni* Poche et *Trypanophis major* Duboscq et Rose. *Archives de Zoologie expérimentale et générale* **74**, 411–435.

Elmhirst, R. (1925). Associations between the amphipod genus *Metopa* and Coelenterates. *Scottish Naturalist* **14**, 149–150.

Fage, L. (1936). Sur un Ellobiopsidé nouveau *Amallocystis fasciatus* gen. et sp. nov. parasite des Mysidacés bathypélagiques. *Archives de Zoologie expérimentale et générale* **78**, N. & R., 145–154.

Fauré-Fremiet, E., Favard, P. and Carasso, N. (1963). Images électroniques d'une microbiocénose marine. *Cahiers de Biologie marine* **4**, 61–64.

Fenaux, R. (1963). Ecologie et biologie des Appendiculaires méditerranéens (Villefranche-sur-Mer). Supplément No. 16 à *Vie et Milieu*, 142 pp.

Field, L. H. (1969). The Biology of *Notophryxus lateralis* (Isopoda, Epicaridea) parasitic on the euphausiid *Nematoscelis difficilis*. *Journal of Parasitology* **55**, 1271–1277.

Fize, A., Manier, J. F. and Maurand, J. (1970). Sur un cas d'infestation du Copépode *Eurytemora velox* (Lillj.) par une levure du genre *Metschnikowia* (Kamenski). *Annales de Parasitologie humaine et comparée* **45**, 357–363.

Flores, M. and Brusca, G. J. (1975). Observations on two species of hyperiid amphipods associated with the ctenophore *Pleurobrachia bachei*. *Bulletin of the Southern California Academy of Science* **74**, 10–15.

Floyd, J. F. (1916). Note on *Trypanophis grobbeni*: a protozoan parasite of Siphonophora. *Proceedings of the Royal Physical Society of Edinburgh* **20**, 62–64.

Franca, S. (1976). On the presence of virus-like particles in the dinoflagellate *Gyrodinium resplendens* (Hulburt). *Protistologica* **12**, 425–430.

Frenzel, J. (1885). Über einige in Seethieren lebende Gregarinen. *Archiv für mikroskopische Anatomie* **24**, 545–588.

Freudenthal, H. D. (1962). *Symbiodinium* gen. nov. and *S. microadriaticum* sp. nov., a zooxanthella; taxonomy, life cycle and morphology. *Journal of Protozoology* **9**, 45–52.

Fuhrmann, O. (1928). Zweite Klasse des Cladus Plathelminthes: Trematoda. *In* "Handbuch der Zoologie" (W. Kükenthal and T. Krumbach eds.), 2, 1, 1–140.

Furnestin, M. L. (1957). Chaetognathes et zooplancton du secteur Atlantique Marocain. *Revue des Travaux de l'Institut des Pêches maritimes* **21**, 1–356.

Furnestin, M. L. and Rebecq, J. (1966). Sur l'ubiquité de *Cercaria owreae* (R. F. Hutton, 1954). *Annales de Parasitologie humaine et comparée* **41**, 61–70.

Galt, J. H. and Whisler, H. C. (1970). Differenciation of flagellated spores in *Thalassomyces* Ellobiopsid parasite of marine Crustacea. *Archiv für Mikrobiologie* **71**, 295–303.

Geiger, S. R. (1969). Distribution and development of mysids (Crustacea Mysidacea) from the Arctic Ocean and confluent seas. *Bulletin of the Southern California Academy of Science* **68**, 103–111.

Gilson, G. (1909). *Prodajus ostendensis* n.sp. Etude monographique d'un épicaride parasite du *Gastrosaccus* spinifer Goes. *Bulletin Scientifique de France et de Belgique* **43**, 19–92.

Gobillard, M. O. (1963). Sur une Grégarine parasite de Copépodes pélagiques. *Vie et Milieu* **14**, 97–105.

Gobillard, M. O. (1964). *Cephaloidophora petiti* sp.n. Grégarine parasite de Copépodes pélagiques de la région de Banyuls (note préliminaire). *Supplément* No. 17 à *Vie et Milieu* (Volume jubilaire, G. Petit), 107–113.

Gönnert, R. (1936). *Sporodinium pseudocalani* n.g.n.sp. ein Parasit aus Copepodeneiern. *Zeitschrift für Parasitologie* **9**, 140–143.

Grabda, E. (1968). Parasite constituents of marine plankton. *Wiadomosci Parazytologicne* **14**, 327–329.

Grassé, P. (1952a). Les Coelomastigina (Chatton et Biecheler, 1935) et les Blastuloidae Neresheimer, 1904). *In* "Traité de Zoologie" (P. Grassé, ed.), Masson, Paris, I, 1020–1022.

Grassé, P. (1952b). Les Ellobiopsidae (Ellobiopsidae Coutière 1911). *In* "Traité de Zoologie" (P. Grassé, ed.), pp. 1023–1030, Masson, Paris.

Grassi, G. B. (1881). Intorno ad alcuni protisti endoparassitici: Flagellati, Lobosi, Sporozoi e Ciliati. *Atti della Società Italiana di Scienze Naturali* **24**, 135–224.

Greeff, R. (1885). Über die pelagische Fauna an den Küsten der Guinea Inseln. *Zeitschrift für wissenschaftliche Zoologie* **42**, 432–458.

Haeckel, E. (1864). Beiträge zur Kenntniss der Corycaeiden. *Jenaische Zeitschrift für Medicin und Naturwissenschaft* **2**, 61–112.

Hall, W. T. and Claus, G. (1963). Ultrastructural studies on the blue-green algal symbiont in *Cyanophora paradoxa* Korschikoff. *Journal of Cell Biology* **19**, 551–563.

Hamon, M. (1951a). Contribution à l'étude cytochimique de *Trypanophis sagittae* Hovasse. *Bulletin scientifique de France et de Belgique* **85**, 176–186.

Hamon, M. (1951b). Note sur une Grégarine parasite du tube digestif de *Sagitta lyra*. *Bulletin de la Société d'Histoire naturelle d'Afrique du Nord* **42**, 11–14.

Hamon, M. (1954). Un nouveau Cilié Hétérotriche, ectoparasite des larves pélagiques de *Luidia*. *Archives de Zoologie expérimentale et générale* **91**, N. & R., 3, 145–156.

Hamon, M. (1957). Note sur *Janickina pigmentifera* (Grassi, 1881) amibe parasite du segment génital mâle de *Sagitta*. *Bulletin de la Société d'Histoire naturelle d'Afrique du Nord* **48**, 220–233.

Hamond, R. (1967). The Amphipoda of Norfolk. *Cahiers de Biologie marine* **8**, 113–152.

Hansen, H. J. (1897). "The Choniostomatidae, a family of Copepoda parasites on Crustacea Malacostracea". A. F. Hyst, Copenhagen, 205 pp.

Harbison, G. R. (1976). Development of *Lycaea pulex* Marion 1874 and *Lycaea vincentii* Stebbing 1888 (Amphipoda, Hyperiidea). *Bulletin of Marine Science* **26**, 152–164.

Harbison, G. R., Biggs, D. C. and Madin, L. P. (1977). Associations of Amphipoda Hyperiidea with gelatinous zooplankton. II. Association with Cnidaria, Ctenophora and Radiolaria. *Deep Sea Research* **24**, 465–488.

Harbison, G. R., Madin, L. P. and Swanberg, N. R. (1978). On the natural history and distribution of oceanic ctenophores. *Deep Sea Research* **25**, 233–256.

Hayashi, K. I. and Miyake, S. (1968). Three Caridean shrimps associated with a medusa from Tanabe Bay. *Publications of the Seto Marine Biological Laboratory* **16**, 11–19.

Hedgpeth, J. W. (1962). A bathypelagic pycnogonid. *Deep Sea Research* **9**, 487–491.

Henon, G. A. and Damkaer, D. M. (1986). New nicothoid Copepod parasitic on Mysids from northwestern North America. *Journal of Crustacean Biology* **6**, 652–665.

Hensen, V. (1887). Über die Bestimmung des Planktons oder des im Meer treibenden Materials an Pflanzen und Thieren; nebst Anhang. *In* "Fünfter Bericht der

Kommission zur wissenschaftliche Untersuchung der deutsche Meere in Kiel für den Jahren 1882–1886", pp. 1–107. Berlin.

Hochberg, F. G. and Seapy, R. R. (1988). The genus *Cephaloidophora* (Apicomplexa: Sporozoa). Septate Gregarine parasites of Pterotracheid Heteropods (Mollusca Gastropoda) (in preparation).

Hoenigman, J. (1958). Sur la découverte de quelques espèces zooplanktoniques nouvelles pour l'Adriatique, de deux épibiontes nouveaux pour les Mysidacés et de deux espèces de parasites nouvelles pour le domaine méditerranéen. *Rapports et Procès-Verbaux des Réunions de la Commission Internationale pour l'Exploration Scientifique de la Mer Méditerranée* **14**, 263–264.

Hoenigman, J. (1960). Faits nouveaux concernant les Mysidacés (Crustacea) et leurs épibiontes dans l'Adriatique. *Rapports et Procès-Verbaux des Réunions de la Commission Internationale pour l'Exploration Scientifique de la Mer Méditerranée* **15**, 339–343.

Hoenigman, J. (1963). Mysidacea de l'Expédition "Hvar" (1948–49) dans l'Adriatique. *Rapports et Procès-Verbaux des Réunions de la Commission Internationale pour l'Exploration Scientifique de la Mer Méditerranée* **17**, 603–616.

Hoenigman, J. (1965). Note sur les Copépodes, leurs épibiontes (Suctoridae) et sur un Mysidacé hôte nouveau de l'*Amallocystis* (Ellobiopsidae). *Rapports et Procès-Verbaux des Réunions de la Commission Internationale pour l'Exploration Scientifique de la Mer Méditerranée* **18**, p. 473.

Hollande, A. (1974). Etude comparée de la mitose syndinienne et de celle des Péridiniens libres et des Hypermastigines. Infrastructure et cycle évolutif des Syndinides parasites de Radiolaires. *Protistologica* **10**, 413–451.

Hollande, A. (1980). Identification du parasome (Nebenkern) de *Janickina pigmenti-fera* à un symbionte (*Perkinsella amoebae* nov.gen.nov.sp.) apparenté aux Fla-gellés Kinétoplastidiés. *Protistologica* **16**, 613–625.

Hollande, A. and Cachon, J. (1952). Un parasite des oeufs de Sardine: l'*Ichthyodinium chabelardi* nov.gen.nov.sp. (Péridinien parasite). *Comptes-rendus de l'Académie des Sciences, Paris* **235**, 976–977.

Hollande, A. and Cachon, J. (1953). Morphologie et évolution d'un Péridinien parasite des oeufs de Sardine (*Ichthyodinium chabelardi*). *Bulletin des Travaux de la Station d'Aquiculture et de Pêche de Castiglione*, N.S. **4**, 319–331.

Hollande, A. and Carré, D. (1974). Les xanthelles des radiolaires Sphaerocollides, des acanthaires et de *Velella velella*: infrastructure–cytochimie–taxonomie. *Protistologica* **10**, 573–601.

Hollande, A. and Corbel, J. C. (1982). Ultrastructure, cycle évolutif et position systématique de *Caryotoma bernardi* Holl. et Enj. (Dinoflagellés Oodinides) parasite endocapsulaire des Thalassicolles (Radiolaires). *Protistologica* **18**, 123–133.

Hollande, A. and Enjumet, M. (1953). Contribution à l'étude biologique des Sphaerocollides (Radiolaires Collodaires et Radiolaires Polycyttaires) et de leurs parasites. *Annales des Sciences Naturelles, Zoologie*, 11e série **15**, 99–183.

Hovasse, R. (1923a). *Endodinium chattoni* (nov.gen.n.sp.) parasite des vélelles. *Bulletin Biologique de France et de Belgique* **57**, 107–130.

Hovasse, R. (1923b). Quelques faits nouveaux concernant le parasitisme de *Blastodinium* et de *Syndinium*. *Comptes-rendus de la Société de Biologie* **89**, 321–322.

Hovasse, R. (1924a). "Zooxanthella chattoni" (*Endodinium chattoni*). Etude complé-mentaire. *Bulletin Biologique de France et de Belgique* **58**, 38–48.

Hovasse, R. (1924b). *Trypanoplasma sagittae* n.sp. *Comptes-rendus de la Société de Biologie* **91**, 1254–1255.

Hovasse, R. (1926). *Parallobiopsis coutieri* Collin. Morphologie, Cytologie, Evolution. Affinités des Ellobiopsidés. *Bulletin biologique de France et de Belgique* **60**, 409–446.

Hovasse, R. (1952). *Ellobiopsis fagei* Hovasse, Ellobiopsidé parasite en Méditerranée de *Clausocalanus arcuicornis* Dana. *Bulletin de l'Institut Océanographique de Monaco* No. **1016**, 1–12.

Hovasse, R. (1974). A propos des Ellobiopsidae. *Actualités Protozoologiques* **1**, 362–363.

Humes, A. G. (1969). A cyclopoid copepod, *Sewellodhiron* n.g.n.sp. associated with a medusa in Porto-Rico. *Beaufortia* **16**, 171–183.

Humes, A. G. (1970). *Paramacrochiron japonicum* n.sp., a cyclopoid copepod associated with a medusa in Japan. *Publications of the Seto Marine Biological Laboratory* **18**, 223–232.

Hutton, R. F. (1954). *Metacercaria owreae* n.sp. an unusual trematode larva from Florida Current chaetognaths. *Bulletin of Marine Science of the Gulf and Caribbean* **4**, 104–109.

Huxley, J. (1910). On *Ganymedes anaspidis* (nov.gen., nov.sp.): a gregarine from the digestive tract of *Anaspides tasmaniae* (Thompson). *Quarterly Journal of Microscopical Science* **55**, 155–175.

Ikeda, I. (1917). A new astomatous ciliate *Metaphrya sagittae* gen. et sp.nov. found in the coelom of *Sagitta*. *Annotationes Zoologicae Japonenses* **9**, 317–324.

Ingold, C. T. (1940). *Endocoenobium eudorinae* gen. et sp.nov., a chytridiaceous fungus parasitizing *Eudorina elegans* Ehrenb. *New Phytologist* **39**, 97–103.

Ingold, C. T. (1944). Studies on British Chytrids. II. A new Chytrid on *Ceratium* and *Peridinium*. *Transactions of the British Mycological Society* **27**, 93–96.

Issi, I. V. (1986). Microsporidia as a phylum of parasitic protozoa (in Russian). *Protozoology* (Academy of Sciences of the USSR) **10**, 6–136.

Janicki, C. (1912a). Untersuchungen an parasitischen Arten der Gattung *Paramoeba* Schaudinn (*P. pigmentifera* Grassi und *P. chaetognathi* Grassi). *Verhandlungen der Naturforschenden Gesellschaft in Basel* **23**, 1–16.

Janicki, C. (1912b). Paramoebenstudien (*P. pigmentifera* Grassi und *P. chaetognathi* Grassi). *Zeitschrift für wissenschaftliche Zoologie* **103**, 449–518.

Janicki, C. (1928). Studien am Genus *Paramoeba* Schaudinn Neue Folge. I Teil. *Zeitschrift für wissenschaftliche Zoologie* **131**, 588–644.

Janicki, C. (1932). Studien am Genus *Paramoeba* Schaudinn Neue Folge. II Teil. *Zeitschrift für wissenschaftliche Zoologie* **142**, 587–623.

Jarling, C. and Kapp, H. (1985). Infestation of Atlantic chaetognaths with helminths and ciliates. *Diseases of Aquatic Organisms* **1**, 23–28.

Jepps, M. W. (1937). On the protozoan parasites of *Calanus finmarchicus* in the Clyde Sea Area. *Quarterly Journal of Microscopical Science* **79**, 589–658.

Jones, G. E. (1958). Attachment of marine bacteria in zooplankton. *United States Fish and Wildlife Service*, Report No. 279, 77–78.

Jones, L. T. (1964). A new host and location for the euphausiid parasite *Thalassomyces fagei* (Boschma) (Protozoa Ellobiopsidae). *Crustaceana* **7**, 148–150.

Kagei, N. (1974). Studies on anisakid Nematoda (Anisakinae). 4. Survey of *Anisakis* larvae in the marine Crustacea. *Bulletin of the Institute of Public Health (Tokyo)* **23**, 65–71.

Kane, J. E. (1964). *Thalassomyces marsupii* a new species of ellobiopsid parasite on the hyperiid amphipod *Parathemisto gaudichaudii* (Guér.). *New Zealand Journal of Science* **7**, 289–303.

Kaneko, T. and Colwell, R. R. (1973). Ecology of *Vibrio parahaemolyticus* in Chesapeake Bay. *Journal of Bacteriology* **113**, 24–32.

Keysselitz, G. (1904). Über *Trypanophis grobbeni* (*Trypanosoma grobbeni* Poche). *Archiv für Protistenkunde* **3**, 367–375.

Kinne, O. (ed.) (1980–1985). "Diseases of Marine Animals", Vol. I, Wiley, New York. Vols II–IV Biologische Anstalt Helgoland, Hamburg.

Koehler, R. (1911). Isopodes nouveaux de la famille des Dajidés provenant des campagnes de la "Princesse Alice". *Bulletin de l'Institut Océanographique de Monaco* **196**, 1–34.

Kofoid, C. A. (1908). The plankton of the Illinois River, 1894–1899 ... Part II. Constituent organisms and their seasonal distribution. *Bulletin of the Illinois State Laboratory of Natural History* **8**, 1–361.

Køie, M. (1975). On the morphology and life-history of *Opechona bacillaris* (Molin 1859) Looss 1907 (Trematoda, Lepocreadiidae). *Ophelia* **13**, 63–86.

Køie, M. (1979). On the morphology and life-history of *Derogenes varicus* (Müller 1784) Looss 1901 (Trematoda, Hemiuridae). *Zeitschrift für Parasitenkunde* **59**, 67–78.

Komaki, Y. (1970). On the parasitic organisms in a krill, *Euphausia similis* from Suruga Bay. *Journal of the Oceanographical Society of Japan* **26**, 283–295.

Kulachkova, V. G. (1972). Helminths from *Sagitta elegans* Verrill from the White Sea (in Russian). *Parazitologya* **6**, 297–304.

Kulka, D. W. and Coney, S. (1984). Incidence of parasitism and irregular development of gonads in *Thysanoessa inermis* (Kroyer) in the Bay of Fundy (Euphausiacea). *Crustaceana* **46**, 87–94.

Labbé, A. (1899). Sporozoa *In* "Das Tierreich", Berlin, 180pp.

Laval, M. (1968). *Zoothamnium pelagicum* Du Plessis, Cilié Péritriche planctonique: morphologie, croissance et comportement. *Protistologica* **4**, 333–363.

Laval, M. (1970). Présence de bactéries intranucléaires chez *Zoothamnium pelagicum* (Cilié Péritriche), leur rôle dans la formation des pigments intracytoplasmiques des zoïdes. *Actes 7e Congrès International de Microscopie électronique, Grenoble* **3**, 403–404.

Laval, M. (1971). Ultrastructure et mode de nutrition du Choanoflagellé *Salpingoeca pelagica* sp.nov., comparaison avec les choanocytes des Spongiaires. *Protistologica* **7**, 325–336.

Laval, M. (1972). Ultrastructure de *Petalotricha ampulla* (Fol) Comparaison avec d'autres Tintinnides et avec les autres ordres de Ciliés. *Protistologica* **8**, 369–386.

Laval, P. (1963). Sur la biologie et les larves de *Vibilia armata* Bov. et de *V. propinqua* Stebb., Amphipodes Hypérides. *Comptes-rendus de l'Académie des Sciences* **257**, 1389–1392.

Laval, P. (1965). Présence d'une période larvaire au début du développement de certains Hypérides parasites (Crustacés Amphipodes). *Comptes-rendus de l'Académie des Sciences, Paris* **260**, 6195–6198.

Laval, P. (1966). *Bougisia ornata* genre et espèce nouveaux de la famille des Hyperiidae (Amphipoda, Hyperiidea). *Crustaceana* **10**, 210–218.

Laval, P. (1968a). Observations sur la biologie de *Phronima curvipes* Voss. (Amphipode Hypéride) et description du mâle adulte. *Cahiers de Biologie marine* **9**, 347–362.

Laval, P. (1968b). Développement en élevage et systématique d'*Hyperia schizogeneios*

Stebb. (Amphipode Hypéride). *Archives de Zoologie expérimentale et générale* **109**, 25–67.

Laval, P. (1972). Comportement, parasitisme et écologie d'*Hyperia schizogeneios* Stebb. (Amphipode Hypéride) dans le plancton de Villefranche-sur-mer. *Annales de l'Institut Océanographique* **48**, 49–74.

Laval, P. (1974a). Un modèle mathématique de l'évitement d'un filet à plancton, son application pratique, et sa vérification indirecte en recourant au parasitisme de l'amphipode Hypéride *Vibilia armata* Bovallius. *Journal of Experimental Marine Biology and Ecology* **14**, 57–87.

Laval, P. (1974b). "Contribution à l'étude des amphipodes Hypérides". Thèse de Doctorat dès Sciences naturelles. Université Paris VI, 80 pp.

Laval, P. (1975). Une analyse multivariable du développement au laboratoire de *Phronima sedentaria* (Forsk.) Amphipode Hypéride. Etude de l'influence de la température et de la quantité de nourriture. *Annales de l'Institut Océanographique* **51**, 5–41.

Laval, P. (1978). The barrel of the pelagic amphipod *Phronima sedentaria* (Forsk.) (Crustacea, Hyperiidea). *Journal of Experimental Marine Biology and Ecology* **33**, 187–211.

Laval, P. (1980). Hyperiid Amphipods as Crustacean parasitoids associated with gelatinous zooplankton. *Oceanography and Marine Biology Annual Review* **18**, 11–56.

Lear, D. W. (1963). Occurrence and significance of chitinoclastic bacteria in pelagic waters and zooplankton. *In* "Marine Microbiology" (C. H. Oppenheimer ed.), pp. 594–610, C. C. Thomas, Springfield.

Lebour, M. V. (1917a). Some parasites of *Sagitta bipunctata*. *Journal of the Marine Biological Association of the United Kingdom* **11**, 201–206.

Lebour, M. V. (1917b). Medusae as hosts for larval trematodes. *Journal of the Marine Biological Association of the United Kingdom* **11**, 57–59.

Lebour, M. V. (1923). Note on the life history of *Hemiurus communis* Odhner. *Parasitology* **15**, 233–235.

Lebour, M. V. (1935). *Hemiurus communis* in *Acartia*. *Journal of the Marine Biological Association of the United Kingdom* **20**, 371–372.

Lee, J. J., Freudenthal, H. D., Kossoy, V. and Bé, A. (1965). Cytological observations on two planktonic Foraminifera, *Globigerina bulloides* d'Orbigny 1826 and *Globigerinoides ruber* (d'Orbigny 1839) Cushman 1927. *Journal of Protozoology* **12**, 531–542.

Leuckart, R. (1860). Bericht über die wissenschaftlichen Leistungen in der Naturgeschichte der niederen Thiere während des Jahres 1859. *Archiv für Naturgeschichte* **26**, 103–264.

Lindley, J. A. (1977). Continuous plankton records: the distribution of the Euphausiacea (Crustacea: Malacostraca) in the north Atlantic and the North Sea, 1966–1967. *Journal of Biogeography* **4**, 121–133.

Lindley, J. A. (1978). Continuous plankton records: the occurrence of apostome ciliates (Protozoa) on Euphausiacea in the North Atlantic Ocean and North Sea. *Marine Biology* **46**, 131–136.

Loeblich, A. R. III (1976). Dinoflagellate evolution: speculation and evidence. *Journal of Protozoology* **23**, 13–28.

Lom, J. (1976). Die "Samtkrankheit"-ein Befall mit Aussenparasiten. *Aquarien Magazin* **10**, 284–287.

Lom, J. (1981). Fish invading dinoflagellates: a synopsis of existing and newly proposed genera. *Folia Parasitologica* **28**, 3–11.

Lom, J. and Lawler, A. R. (1973). An ultrastructural study of the mode of attachment in dinoflagellates invading gills of Cyprinodontidae. *Protistologica* **9**, 293–309.

Mc Laughlin, J. J. A. and Zahl, P. A. (1959). Axenic zooxanthellae from various invertebrate hosts. *Annals of the New York Academy of Science* **77**, 55–72.

Mc Laughlin, J. J. A. and Zahl, P. A. (1966). Endozoic Algae, *In* "Symbiosis" (S. M. Henry, ed.), I, 257–297.

Madin, L. P. and Harbison, G. R. (1977). The associations between Amphipoda Hyperiidea and gelatinous zooplankton. I. Associations with Salpidae. *Deep Sea Research* **24**, 449–463.

Makings, P. (1981). *Mesopodopsis slabberi* (Mysidacea) at Millport, W. Scotland with the parasitic Nematode *Anisakis simplex. Crustaceana* **41**, 310–312.

Margolis, L. (1971). Polychaetes as intermediate hosts of helminth parasites of vertebrates: a review. *Journal of the Fisheries Research Board of Canada* **28**, 1385–1392.

Marquès, A. (1984). "Contribution à la connaissance des Actinomyxidies: ultrastructure, cycle biologique, systématique". Thèse Doctorat des Sciences, Université de Montpellier, 218 pp.

Marquès, A. and Ormières, R. (1981). Adaptation à la vie planctonique des spores d'Actinomyxidies d'après étude en microscopie électronique à balayage. *Comptes-rendus de l'Académie des Sciences, Paris* **293**, 287–288.

Marquès, A. and Ormières, R. (1982). La spore des Actinomyxidies: *Synactinomyxon longicauda* n.sp. un nouveau type de groupement sporal et adaptations planctoniques. *Journal of Protozoology* **29**, 195–202.

Martin, R. and Brinckmann, A. (1963). Zum Brutparasitismus von *Phylirrhoe bucephala* Pér. & Les. auf der Meduse *Zanclea costata* Gegenb. *Publicazione della Stazione Zoologica di Napoli* **33**, 206–223.

Masi, L. (1905). Sulla presenza della *Meganyctiphanes norvegica* (M. Sars) nelle Acque del Giglio. *Annali Museo Civico di Storia Naturale G. Doria*, Ser. 3 a, **2**, 149–156.

Mauchline, J. (1966). *Thalassomyces fagei*, an ellobiopsid parasite of the euphausiid crustacean *Thysanoessa raschi. Journal of the Marine Biological Association of the United Kingdom* **46**, 531–539.

Mauchline, J. (1968). The biology of *Erythrops serrata* and *E. elegans* (Crustacea, Mysidacea). *Journal of the Marine Biological Association of the United Kingdom* **48**, 455–464.

Mauchline, J. (1969). Choniostomatid parasites on species of *Erythrops* (Crustacea Mysidacea). *Journal of the Marine Biological Association of the United Kingdom* **49**, 391–392.

Mauchline, J. (1970). The biology of *Mysidopsis gibbosa, M. didelphys* and *M. angusta* (Crustacea Mysidacea). *Journal of the Marine Biological Association of the United Kingdom* **50**, 381–396.

Mauchline, J. (1971). Crustacea Mysidacea *In* "Fauna of the Clyde Sea Area" (Scottish Marine Biological Association), 1–26.

Mauchline, J. (1980). The biology of mysids and euphausiids. *Advances in Marine Biology* **18**, 681 pp.

Mayer, J. A. and Taylor, F. J. R. (1979). A virus which lyses the marine nanoflagellate *Micromonas pusilla, Nature* **281**, 299–301.

Menon, M. G. K. (1930). The Scyphomedusae of Madras and neighbouring coast. *Bulletin of the Madras Government Museum (Natural History)* **3**, 1–28.

Mingazzini, P. (1891). Gregarine monocistidae nuove o poco conosciute, del Golfo di Napoli. *Rendiconti dell'Academia di Lincei* **4**, 467–474.

Mingazzini, P. (1893). Contributo alla conoscenza degli Sporozoi. *Ricerche fatte nel Laboratorio di Anatomia normale della Reale Università di Roma* **3**, 31–85.

Morris, E. O. (1968). Yeasts of marine origin. *Oceanography and Marine Biology Annual Review* **6**, 20–30.

Nagasawa, S. and Marumo, R. (1979). Parasites of Chaetognaths in Suruga Bay (Japan). *La Mer (Tokyo)* **17**, 127–136.

Nicol, S. (1984). *Ephelota* sp., a suctorian found on the euphausiid *Meganyctiphanes norvegica*. *Canadian Journal of Zoology* **62**, 744–746.

Norris, R. E. (1967). Algal consortisms in marine plankton. In "Proceedings of a Seminar on Sea, Salt and Plants" (V. Krishnamurty, ed.), 178–189.

Nouvel, H. (1941). Sur les Ellobiopsidés des Mysidacés provenant des Campagnes du Prince de Monaco. *Bulletin de l'Institut Océanographique* **809**, 1–8.

Nouvel, H. (1951). *Gastrosaccus normani* G. O. Sars 1877 et *Gastrosaccus lobatus* n.sp. (Crust. Mysid.) avec précision de l'hôte de *Prodajus lobiancoi* Bonnier (Crust. Isop. Epicar.) *Bulletin de l'Institut Océanographique de Monaco* **993**, 1–12.

Nouvel, H. and Hoenigman, J. (1955). *Amallocystis boschmai* Nouvel 1954 Ellobiopsidé parasite du Mysidacé *Leptomysis gracilis* G. O. Sars. *Travaux du Comité local d'Océanographie et d'Etude des Côtes de l'Algérie*, No. 26, Campagne du "Professeur Lacaze-Duthiers" sur les côtes d'Algérie, juin-juillet 1952, 7–18.

Øresland, V. (1986). Parasites of the chaetognath *Sagitta setosa* in the western English Channel, *Marine Biology* **92**, 87–91.

Ormières, R. (1965). Recherches sur les Sporozoaires parasites des Tuniciers. *Vie et Milieu* **15**, 823–846.

Oshima, T., Shimazu, T., Koyama, H. and Akahane, H. (1969). On the larvae of the genus *Anisakis* (Nematoda Anisakinae) from the euphausiids. *Japanese Journal of Parasitology* **18**, 241–248.

Pearre, S. (1976). Gigantism and partial parasitic castration of Chaetognatha infected with larval trematodes. *Journal of the Marine Biological Association of the United Kingdom* **56**, 503–513.

Pearre, S. (1979). Niche modification in Chaetognatha infected with larval trematodes (Digenea). *Internationale Revue der Gesamten Hydrobiologie* **64**, 193–206.

Pillai, N. K. (1963). Isopod parasites of Indian mysids. *Annals and Magazine of Natural History*, Ser. 13, **6**, 739–743.

Poche, F. (1903). Über zwei neue in Siphonophoren vorkommende Flagellaten nebst Bemerkungen über die Nomenclatur einiger verwandter Formen. *Arbeiten aus dem Zoologische Institute der Universität Wien* **14**, 307–358.

Rao, H. and Madhavi, R. (1966). Tetraphyllidean larva (Cestoda) in the Copepod *Eucalanus subcrassus* Giesbrecht 1888 off Waltair Coast, Bay of Bengal. *Current Science* **35**, 70–71.

Ramirez, F. C. and Dato, C. (1989). Observations on parasitism by *Thalassomyces fagei* of three euphausiid species in Southern waters. *Oceanologica Acta* **19**, 95–97.

Rebecq, J. (1965). Considérations sur la place des trématodes dans le zooplancton marin. *Annales de la Faculté des Sciences de l'Université d'Aix-Marseille* **38**, 61–84.

Reddiah, K. (1968). Three new species of *Paramacrochiron* (Lichomolgidae) associated with medusae. *Crustaceana, Supplement* I, 193–209.

Reddiah, K. (1969). *Pseudomacrochiron stocki* n.g.n.sp. a cyclopoid associated with a medusa. *Crustaceana* **16**, 43–50.

Reimer, L. W. (1976). Metacercariae in invertebrates of the Madras coast. *Angewandte Parasitologie* **17**, 33–43.

Reimer, L. W. (1977). Larval Cestodes in planktonic invertebrates of the Atlantic near the coast of north-west Africa. *Parazitologiya* **11**, 309–315.

Reimer, L. W., Hnatiuk, S. and Rochner, J. (1975). Metacercarien in Planktontieren des mittleren Atlantik. *Wissenschaftliche Zeitschrift der pädagogische Hochschule Güstrow* **2**, 239–258.

Roboz, Z. (1886). Adatok a gregarinak Ismereténez. *Ertekezések Termész Magyar Akademia* **16**, 1–34.

Rose, M. (1933a). *Monocystis copiliae* n.sp. Grégarine parasite d'un Copépode pélagique: *Copilia vitraea* Haeckel. *Bulletin de la Société d'Histoire Naturelle d'Afrique du Nord* **24**, 357–359.

Rose, M. (1933b). *Nogagella* n.g. *siphonophoriae* n.sp. Copépode Caligide parasite des Siphonophores. *Annales de l'Institut Océanographique* **13**, 119–133.

Rose, M. (1936). Documents pour servir à l'étude des infusoires Ciliés Apostomes parasites des Siphonophores et des organismes pélagiques. *Archives de Zoologie expérimentale et générale* **78**, N. & R., 184–198.

Rose, M. (1939). Sur le passage de la forme flagellée à la forme grégarinienne chez *Trypanophis grobbeni* Poche. *Archives de Zoologie expérimentale et générale* **80**, Notes et Revue, 39–48.

Rose, M. (1947). Sur le cycle évolutif de *Trypanophis grobbeni* Poche. *Bulletin Biologique de France et de Belgique* **81**, 6–32.

Rose, M. and Cachon, J. (1951). *Diplomorpha paradoxa* nov.gen. nov.sp. protiste de l'ectoderme des Siphonophores. *Comptes-rendus de l'Académie des Sciences* **233**, 451–452.

Rose, M. and Cachon, J. (1952a). Le mouvement chez *Diplomorpha paradoxa* Rose et Cachon, parasite de Siphonophores. *Comptes-rendus de l'Académie des Sciences, Paris* **234**, 669–671.

Rose, M. and Cachon, J. (1952b). L'émission des bras chez *Diplomorpha paradoxa* Rose et Cachon, protiste parasite de Siphonophores. *Comptes-rendus de l'Académie des Sciences, Paris* **234**, 2306–2308.

Rose, M. and Hamon, M. (1950). Une nouvelle espèce de *Trypanophis, T. sagittae* Hovasse 1924. *Bulletin Biologique de France et de Belgique* **84**, 101–115.

Rose, M. and Hamon, M. (1952). Sur un Copépode parasite de certains Mollusques Ptéropodes Thécosomes. *Annales des Sciences Naturelles, Zoologie*, 11e Série, 219–230.

Sars, G. O. (1885). Report on the scientific results of the voyage of H.M.S. "Challenger" during the years 1873–1876. *Zoology* **13** (37), 1–228.

Sars, M. (1868). Bidrag til kundskab om Christianiafjordens Fauna. *Nyt Magazin for Naturvidenskaberne* **15**, 1–104.

Scott, D. M. (1957). Records of larval *Contracaecum* sp. in three species of mysids from the Bras d'Or Lakes, Nova Scotia, Canada. *Journal of Parasitology* **43**, 290.

Sebastian, M. J. (1970). On two isopod parasites of Indian euphausiids. *Journal of Natural History* **4**, 153–158.

Seki, H. and Fulton, J. (1969). Infection of marine copepods by *Metschnikowia* sp. *Mycopathologia et Mycologia Applicata* **38**, 61–70.

Selensky, W. D. (1927). On a new Ichthyobdellidae (*Mysidobdella oculata* n.g.n.sp.) a parasite of Crustacea. *Proceedings of the 2nd Congress of Zoology, Anatomy and Histology of USSR* **2**, 32–33 (in Russian).

Sentz-Braconnot, E. and Carré, C. (1966). Sur la biologie du Nudibranche pélagique *Cephalopyge trematoides*. Parasitisme sur le Siphonophore *Nanomia bijuga*, nutrition, développement. *Cahiers de Biologie marine* **7**, 31–38.

Sewell, R. B. S. (1951). The epibionts and parasites of the planktonic Copepoda of the Arabian Sea. *John Murray Expedition Science Reports, British Museum (Natural History)* **8**, 317–592.

Sherman, K. and Schaner, E. G. (1965). *Paracineta* sp. an epizoic Suctorian found on Gulf of Maine Copepoda. *Journal of Protozoology* **12**, 618–625.

Shimazu, T. (1975a). On the parasitic organisms in a krill, *Euphausia similis* from Suruga Bay. V. Larval Cestodes. *Japanese Journal of Parasitology* **24**, 122–128.

Shimazu, T. (1975b). Some cestode and acanthocephalan larvae from euphausiid crustaceans collected in the northern North-Pacific Ocean. *Bulletin of the Japanese Society of Scientific Fisheries* **41**, 813–821.

Shimazu, T. (1978). Some helminth parasites of the Chaetognatha from Suruga Bay, Central Japan. *Bulletin of the National Science Museum of Tokyo*, Ser. A. *(Zoology)* **4**, 105–116.

Shimazu, T. (1979). Some protozoan parasites of the Chaetognatha from Suruga Bay, Central Japan. *Japanese Journal of Parasitology* **28**, 51–55.

Shimazu, T. (1981). Metacercariae of *Neonotoporus trachuri* (Trematoda, Lepocreadiidae) parasitic to *Euphausia similis* (Crustacea, Euphausiidae). *Zoological Magazine* **90**, 254–257.

Shimazu, T. (1982). Some helminth parasites of marine planktonic invertebrates. *Journal of Naganoken Junior College* **32**, 11–29.

Shimazu, T. and Oshima, T. (1972). Some larval nematodes from euphausiid crustaceans. *In* "Biological Oceanography of the Northern North Pacific Ocean" (A. Takenouti *et al.*, eds.), pp. 403–409. Indemitsu Shojen, Tokyo.

Sicko-Goad, L. and Walker, G. (1979). Viroplasm and large virus-like particles in the dinoflagellate *Gymnodinium uberrimum*. *Protoplasma* **99**, 203–210.

Sieburth, J. Mc N. (1979). "Sea microbes", 491 pp. Oxford University Press, New York.

Silva, E. S. (1967). *Cochlodinium heterolobatum* n.sp. structure and some cytophysiological aspects. *Journal of Protozoology* **14**, 745–754.

Silva, E. S. (1978). Endonuclear bacteria in two species of dinoflagellates. *Protistologica* **14**, 113–119.

Silva, E. S. and Franca, S. (1985). The association dinoflagellate-bacteria: their ultrastructural relationship in two species of dinoflagellates. *Protistologica* **21**, 429–446.

Simidu, U., Ashino, K. and Kaneko, E. (1971). Bacterial flora of phyto- and zooplankton in the inshore water of Japan. *Canadian Journal of Microbiology* **17**, 1157–1160.

Slankis, A. J. and Shevehenko, G. G. (1974). Some data on the infestation of the plankton invertebrates with helminth larvae in the west part of the tropical zone of the Pacific. *Transactions of the Pacific Research Institute of Fisheries and Oceanography (Vladivostok)* **88**, 129–138 (in Russian).

Smith, J. W. (1971). *Thysanoessa inermis* and *T. longicaudata* (Euphausiidae) as first intermediate hosts of *Anisakis* sp. (Nematoda Ascaridata) in the northern North Sea, to the north of Scotland and at Faroe. *Nature* **234**, 478.

Sournia, A., Cachon, J. and Cachon, M. (1975). Catalogue des espèces et taxons infraspécifiques de Dinoflagellés marins actuels publiés depuis la revision de J. Schiller. II. Dinoflagellés parasites ou symbiotiques. *Archiv für Protistenkunde* **117**, 1–19.

Soyer, M. O. (1965). Une nouvelle Eugrégarine parasite de Sapphirinidae (Copepoda Podoplea). *Vie et Milieu* **16**, 243–249.

Soyer, M. O. (1974). Etude ultrastructurale de *Syndinium* sp. Chatton, parasite coelomique de Copépodes pélagiques. *Vie et Milieu* **24**, fasc. 2, A, 191–212.

Soyer, M. O. (1978). Particules de type viral et filaments trichocystoides chez les Dinoflagellés. *Protistologica* **14**, 53–58.

Sprague, V. (1965). *Ichthyosporidium* Caullery and Mesnil 1905, the name of a genus of fungi or a genus of Sporozoans? *Systematic Zoology* **14**, 110–114.

Sprague, V. (1966). *Ichthyosporidium* sp. Schwartz, 1963, parasite of the fish *Leiostomus xanthurus*, is a microsporidian. *Journal of Protozoology* **13**, 356–358.

Sprague, V. and Hussey, K. L. (1980). Observations on *Ichthyosporidium giganteum* (Microsporida) with particular reference to the host-parasite relations during merogony. *Journal of Protozoology* **27**, 169–175.

Sprague, V. and Vernick, S. H. (1974). Fine structure of the cyst and some sporulations stages of *Ichthyosporidium* (Microsporida). *Journal of Protozoology* **21**, 667–677.

Sproston, N. G. (1944). *Ichthyosporidium hoferi* (Plehn and Mulsow, 1911) an internal fungoid parasite of the mackerel. *Journal of the Marine Biological Association of the United Kingdom* **26**, 72–98.

Steuer, A. (1928). On the geographical distribution and affinity of the appendiculate trematodes parasitizing marine plankton copepods. *Journal of Parasitology* **15**, 115–120.

Stock, J. (1971). *Micrallecto uncinata* n.gen.n.sp. a parasitic copepod from a remarkable host, the Pteropod *Pneumoderma*. *Bulletin of the Zoological Museum of the University of Amsterdam* **2**, 77–81.

Stock, J. (1973). *Nannallecto fusii* n.gen.n.sp. a copepod parasitic in the pteropod *Pneumodermopsis*. *Bulletin of the Zoological Museum of the University of Amsterdam* **3**, 21–24.

Stock, J. and Van der Spoel, S. (1976). *Pteroxena papillifera* n.gen.n.sp., an endoparasitic organism (Copepoda?) from the gymnosomatous pteropod *Notobranchaea*. *Bulletin of the Zoological Museum of the University of Amsterdam* **5**, 177–180.

Stuart, A. (1871). Über den Bau der Gregarinen. *Bulletin de l'Académie Impériale des Sciences de Saint Pétersbourg* **15**, 497–502.

Subrahmanian, R. (1954). A new member of the Euglenineae, *Protoeuglena noctilucae* gen. et sp.nov. occurring in *Noctiluca miliaris* Suriray, causing green discoloration of the sea off Calicut. *Proceedings of the Indian Academy of Science* **39B**, 118–127.

Swanberg, N. R. and Harbison, G. R. (1980). The ecology of *Collozoum longiforme* new species, a new colonial radiolarian from the equatorial Atlantic Ocean. *Deep Sea Research*, Pt. A. **27**, 715–732.

Sweeney, B. M. (1976). *Pedinomonas noctilucae* (Prasinophyceae), the flagellate symbiotic in *Noctiluca* in Southeast Asia. *Journal of Phycology* **12**, 460–464.

Taberly, G. (1954). Etude morphologique d'un Dajidae peu connu: *Prodajus lobiancoi* Bonnier (Crust. Isop. Epicaride). I. L'epicaridium de *P. lobiancoi* et remarques générales sur l'epicaridium des Dajidae. *Bulletin de l'Institut Océanographique de Monaco* **1045**, 1–9.

Tattersall, W. M. and Tattersall, O. S. (1951). The British Mysidacea, 460pp. Ray Society Monograph.
Taylor, D. L. (1971). Ultrastructure of the "zooxanthella" *Endodinium chattonii in situ*. *Journal of the Marine Biological Association of the United Kingdom* **51**, 227–234.
Taylor, D. L. (1973). The cellular interactions of algal-invertebrate symbioses. *Advances in Marine Biology* **11**, 1–56.
Taylor, F. J. R. (1968). Parasitism of the toxin-producing dinoflagellate *Gonyaulax catenella* by the endoparasitic dinoflagellate *Amoebophrya ceratii*. *Journal of the Fisheries Research Board of Canada* **25**, 2241–2245.
Théodoridès, J. and Carré, C. (1969). Parasitisme d'Alciopidae (Annelida Polychaeta) de Villefranche-sur-mer par des Grégarines (Eugregarina). *Annales de Parasitologie humaine et comparée* **44**, 519–520.
Théodoridès, J. and Desportes, I. (1968). Sur trois Grégarines parasites d'Invertébrés marins. *Bulletin de l'Institut Océanographique de Monaco* **1387**, 1–11.
Théodoridès, J. and Desportes, I. (1972). Mise en évidence de nouveaux représentants de la famille des Ganymedidae Huxley, Grégarines parasites de Crustacés. *Comptes-rendus de l'Académie des Sciences* **274**, 3251–3253.
Théodoridès, J. and Desportes, I. (1975). Sporozoaires d'Invertébrés pélagiques de Villefranche-sur-mer (étude descriptive et faunistique). *Protistologica* **11**, 205–220.
Thiel, M. E. (1976). Wirbellose Meerestiere als Parasiten, Kommensalen oder Symbionten in oder an Scyphomedusen. *Helgoländer wissenschaftliche Meeresuntersuchungen* **28**, 417–446.
Trégouboff, G. (1918). Etude monographique de *Gonospora testiculi* Trég., Grégarine parasite du testicule de *Cerithium vulgatum* Burg. *Archives de Zoologie expérimentale et générale* **57**, 471–509.
Trégouboff, G. and Rose, M. (1957). "Manuel de Planctonologie méditerranéenne", 2 vols. 587 pp + 207 pl., Centre National de la Recherche Scientifique, Paris.
Trench, R. K. (1987). Dinoflagellates in non-parasitic symbioses, *In* "The Biology of Dinoflagellates" (F. J. R. Taylor, ed.), pp. 530–570, Blackwell, Oxford.
Vader, W. (1972a). Associations between amphipods and molluscs. A review of published records. *Sarsia* **48**, 13–18.
Vader, W. (1972b). Associations between gammarid and caprellid amphipods and medusae. *Sarsia* **50**, 51–56.
Vader, W. (1973a). The oldest published record of a *Thalassomyces* species (Ellobiopsidae). *Sarsia* **52**, 171–174.
Vader, W. (1973b). A bibliography of the Ellobiopsidae, 1959–1971, with a list of *Thalassomyces* species and their hosts. *Sarsia* **52**, 175–180.
Vader, W. and Kane, J. E. (1968). New hosts and distribution record of *Thalassomyces marsupii* Kane, an ellobiopsid parasite on amphipods. *Sarsia* **33**, 13–20.
Walter, E. D., Valovaya, M. A. and Popova, T. I. (1979). A study of the prevalence of helminths in plankton invertebrates in the White Sea (USSR) (in Russian). *Vestnik Moskovskogo Universiteta*, Series 16, *Biologiya* **3**, 31–38.
Wayne Coats, D. (1988). *Duboscquelle cachoni* n. sp. a parasitic dinoflagellate lethal to its Tintinnine host *Eutinitinnus pectinis*. *Journal of Protozoology* **35**, 607–617.
Weill, R. (1935). Revue des Protistes commensaux ou parasites des Cnidaires. *Archives de Zoologie expérimentale et générale* **77**, N. & R., 2, 47–70.
Weinstein, M. (1972). Studies on the relationship between *Sagitta elegans* Verrill and its endoparasites in the southwestern Gulf of St Lawrence. Ph. D. Thesis, McGill University, Montreal, XI + 202 pp.

White, M. G. and Bone, D. G. (1972). The interrelationship of *Hyperia galba* (Crustacea Amphipoda) and *Desmonema gaudichaudi* (Scyphomedusae Semeostomae) from the Antarctic. *Bulletin of the British Antarctic Survey* **27**, 39–49.

Wing, B. L. (1966). *Thalassomyces* sp. (Ellobiopsidae) infesting *Acanthomysis pseudomacropsis* and *Neomysis kadiakensis* (Mysidacea) in southeastern Alaska. *Deep Sea Research* **13**, 1385.

Wing, B. L. (1975). New records of Ellobiopsidae (Protista *incertae sedis*) from the North Pacific with a description of *Thalassomyces albatrossi* n.sp. a parasite of the mysid *Stylomysis major*. *Fishery Bulletin of the United States National Marine Fisheries* **73**, 169–185.

Wundsch, H. H. (1912). Neue Plerocercoide aus marinen Copepoden. *Archiv für Naturgeschichte*, 78, *Abteilung* A, 1–20.

Yip, S. Y. (1984). Parasites of *Pleurobrachia pileus* Mueller 1776 (Ctenophora) from Galway Bay, western Ireland. *Journal of Plankton Research* **6**, 107–121.

Sandy-Beach Bivalves and Gastropods: A Comparison Between *Donax serra* and *Bullia digitalis*

A. C. Brown, J. M. E. Stenton-Dozey and E. R. Trueman

Department of Zoology, University of Cape Town, Rondebosch 7700, South Africa

ADVANCES IN MARINE BIOLOGY,
VOLUME 25 ISBN 0–12–026125–1

I. Introduction

On clean, high-energy sandy beaches, the macrofauna tends to be dominated by filter-feeders and carnivorous scavengers. On the west and south coasts of southern Africa, both these niches are filled by molluscs—the burrowing, suspension-feeding bivalve *Donax serra* Röding and scavenging nassariid whelks of the genus *Bullia*. The only important intertidal species of the latter genus on the west coast of South Africa is *Bullia digitalis* (Dillwyn), Fig. 1, but on the south coast its distribution overlaps those of *Bullia rhodostoma* Reeve and *Bullia pura* Melville. Similarly, *Donax serra* (Fig. 2) is the only *Donax* species found intertidally on the west coast, while on the south coast it is joined by *Donax sordidus* Hanley. Other species of both *Bullia* and *Donax* occur subtidally on both coasts.

B. *digitalis* and *D. serra* have very similar geographical ranges, extending from Namibia, on the west coast, round the Cape to the Transkei in the south-east (Day, 1974; Kilburn and Rippey, 1982)—a span of some 2000 km. They also favour the same beaches, being rare or absent where wave action is minimal or particularly intense, or where piles of kelp are commonly washed ashore, and reach their greatest densities on clean, relatively gently-sloping beaches of moderate to high wave energy not subject to violent rip or long-shore currents. Not only do the two animals occur on the same beaches but, at least during low water, typically occupy the same zone, the centres of their populations being near the lower tidal limit, although both may move further off-shore under certain conditions. They do, however, display some vertical separation in the sand column, *Bullia* being a much shallower burrower than *Donax* and, unlike *Donax*, emerging regularly from the sand in search of food.

Both species thus face the same conditions, the same unstable substratum, the same harsh environment, and both are very successful, as indicated by their wide distribution and high densities, particularly on those beaches not seriously exploited by man. On such beaches they may together constitute

FIG. 1. (A) The whelk *Bullia digitalis* commencing burrowing (shell length 50 mm).
(B) *Bullia digitalis* surfing (shell length 45 mm). The animal is swept along by the wash, holding its foot rigid as an underwater sail.

well over 95% of the total intertidal biomass. McLachlan (1977a,b) has demonstrated that on high energy sandy beaches along the northern shores of St Francis Bay and Algoa Bay, *D. serra* alone constitutes from 80 to 98% of the total macrofaunal biomass and Bally (1981) indicates a similar situation in some areas of the west coast. Maximum densities encountered by Bally were 1748 individuals of *D. serra* per square metre and 204 individuals of *B. digitalis*. Brown (unpublished) has counted over 300 *B. digitalis*/m^2 on M'Nandi Beach in False Bay but most of these were small juveniles.

The sizes of both molluscs facilitate experimental work; indeed *D. serra* is the largest member of its genus, attaining a shell length of 80 mm and occasionally up to 88 mm, while *B. digitalis* reaches a shell length of 60 mm in very old individuals (Kilburn and Rippey, 1982). Both are also easy to maintain in the laboratory. These and other considerations have led to much research being done on them, so that we considered it appropriate at this stage to attempt some comparison between them and to review their adaptations to a common environment. Adaptations to be considered include activities such as burrowing and surfing, maintenance of position on the shore, avoidance of predators and other features that allow the animals to survive, grow and reproduce in an unstable environment dominated by an erratic food supply and the absence of attached plants. Respiratory and

FIG. 2. (A) *Donax serra* burrowing (shell length 40 mm). The foot is anchored in the sand and the shell is about to be pulled upright. (B) Posterior view of a *Donax serra* half buried in sand. The intricate complex of papillae guarding the entrance to the inhalent siphon (towards the camera) can be seen and behind it the partly extended exhalent siphon.

energetic considerations would appear to be central and it is on these and on locomotory adaptations that the present review concentrates.

There are even more fundamental comparisons to be made than those mentioned above, for prosobranch whelks are essentially epifaunal in habit and secondary adaptation to the infaunal habit in *Bullia* contrasts with the primitively infaunal nature of the Bivalvia (Runnegar and Pojeta, 1985). Thus comparison of the two species, one from each of these taxa, living in a common substratum, provides an interesting study in convergent adaptation.

II. The Sandy Beach Environment

Sandy beaches constitute some 42.5% of the South African coastline (i.e. between the Orange River on the west coast and Kosi Bay on the east) (Underhill and Cooper, 1982; Bally *et al.*, 1984). This figure does not include shores of mixed sand and rock, some of which are partially colonized by sandy-beach animals, including *Donax* and *Bullia*. Exposure to wave action is intense by European standards; the Atlantic swell may reach a height of 18 m and along the south-west coast waves exceeding 6 m occur for some 10% of the time (Shillington, 1978). McLachlan (1980a) has developed an internationally applicable scale for exposure to wave action, ranging from 1 to 20; on this scale almost all South African beaches, except those in estuaries and lagoons, rate between 12 (exposed) and 17 (very exposed).

As a result of this wave action, the amount of fine sand in the South African oceanic beach sediments is low, median grain size on the open shore ranging from 220 μm to 1000 μm as wave action increases (Field and Griffiths, 1989). Pebble beaches are, however, rare, comprising well under 1% of the coastline, and muddy sands are limited to estuaries and sheltered lagoons. In the area under discussion (Namibia to Transkei), the open sandy beaches consist of clean, white sand, made up predominantly of quartzite, with a much lower content of calcium carbonate. They tend to have a moderate to steep slope, high porosity, good drainage and full oxygenation to a depth of 1 m or more below the sand surface (Brown, 1971a; Bally, 1981; Field and Griffiths, 1989). Although in general the more exposed the beach the greater its slope, of more importance to the fauna may be the fact that stability decreases as wave action increases, leading to more and more drastic (and unpredictable) changes in beach profile (Brown, 1971a). South African beaches are mostly too unstable to support the burrows of macrofaunal species intertidally and the infauna may be expected to be smothered with sand or washed away together with its substratum, without warning. Tides and wave action combine in the formation of bars and troughs, which in turn influence the patchy distribution of both *Donax* (Stenton-Dozey, 1989) and *Bullia* (Brown, unpublished).

A factor which is commonly neglected, but of considerable importance to all burrowing macrofauna, is the penetrability of the sand. This has been studied on a west coast beach by Bally (1983a) and for Cape Peninsula beaches by Brown and Trueman (unpublished). While the force required to penetrate the sand surface shows a considerable range and can exceed 90 N/ cm², resistance in the wash zone, where molluscs usually burrow, seldom requires an instantaneous force greater than 20 N and this is reduced to 1 N/ cm² or less by gentle sustained pressure or by probing. Resistance decreases with increasing angle from the vertical; at an angle of about 10°C off horizontal, comparable with the angle adopted by *Bullia* in burrowing, resistance is an order of magnitude less than that encountered vertically. Resistance increases rapidly with depth below the surface.

The region displays a simple semi-diurnal tidal regime, having a spring-tide range of 2–2.5 m and a neap-tide amplitude of about 1 m. Spring-tide lows occur between 8 and 10 o'clock, morning and evening, so that the animals living low down the intertidal slope are never exposed to the noon-day sun and experience little heat stress or desiccation.

Water temperatures along the coast are by no means uniform. The south coast is strongly influenced by the warm Agulhas current, which flows in a south-westerly direction, so that this stretch of coast is commonly described as a "warm-temperate region" (Brown and Jarman, 1978). In contrast, the west coast is influenced by the north-flowing, cold Benguela current (Hart and Currie, 1960). This region is usually referred to as being "cold-temperate" (Stephenson and Stephenson, 1972; Brown and Jarman, 1978), although Ekman (1953) and Briggs (1974) considered it to be warm temper-ate. However that may be, these west coast waters are the coldest of the entire African coastline (Brown and Jarman, 1978).

Mean monthly sea surface temperatures on the south coast range from 15° in winter to 22°C in summer (Field and Griffiths, 1989). On the west coast, temperatures vary from 8 to 18°C but not on a seasonal basis. Indeed, there is relatively little difference between the mean winter and summer temperatures on this coast, the marked fluctuations which occur being short-term and due to intermittent upwelling, particularly during summer (Andrews, 1974). Comparatively warm coastal water is blown away from the shore by the south-easterly wind of summer, to be replaced by cold, upwelling water. As the wind is erratic, so is the upwelling it generates, with the result that surface temperatures vary greatly, not only from day to day but even on occasion from hour to hour, in contrast to the relatively gentle rise and fall in temperature experienced along most of the south coast. Indeed, a temperature change from 17 to 8°C has been recorded on the west coast over a single 7-hour period (Branch and Branch, 1981).

From time to time both the west and south coasts may be subject to unusually high water temperatures, such as occurred during the "warm event" of the summer of 1982/83. Branch (1984) has provided evidence of the impact of this event on intertidal and shallow-water communities, while Birkett and Cook (1987) have described the effect of the anomaly on the breeding cycle of *D. serra*.

Although current patterns vary considerably from beach to beach, those on the south coast commonly display cellular circulation patterns in the surf zone, resulting in semi-closed ecosystems which tend to retain nutrients leaching from the beach. This in turn leads to high densities of surf-zone phytoplankton which become trapped close in-shore (McLachlan, 1981; McLachlan and Lewin, 1981). *Anaulus birostratus* is the common diatom in these communities, although other species, such as *Aulacodiscus kittoni*, also occur. On such beaches, primary production within 500 m of the shore has been estimated at 99 kg C/m/y. (McLachlan and Bate, 1984). These rich diatom communities support a varied and quite dense community of surf-zone animals, some of which are eventually eaten by *Bullia*, while the diatoms themselves represent food for *Donax* and other filter feeders.

Circulating cells and their attendant dense phytoplankton are far less common on the west coast, due to the intermittent but strong offshore movement of surface water. In fact, during west coast upwelling, there may be virtually no phytoplankton in the surf, although the nutrient-rich upwelled water continues to promote phytoplankton development off-shore. Should the wind change, leading to downwelling, this phytoplankton may be swept ashore; however by the time it reaches the shore it is usually in a state of decay and arrives as foam (Stenton-Dozey, 1989). Also due to the upwelling regime, the supply of carrion to the beach is even more erratic on the west coast than it is on the south.

A further difference between the two coasts is the far greater input of kelp detritus on the west coast (Bally, 1987). This input, derived mainly from *Ecklonia maxima* and *Laminaria pallida*, varies considerably from beach to beach, being highest near off-shore, rocky platforms dominated by kelp. It arrives in the intertidal zone in sizes that range from the microscopic to complete kelp plants, 2 m or more in length. The microscopic particles, and the bacteria associated with them, may form valuable food for *Donax* but where large quantities of semi-intact kelp wash up onto the beach, both *Donax* and *Bullia* are prevented from colonizing the sand, as the surging back and forth of these detached plants displaces them and seriously disrupts their behaviour.

With the exception of a few impact areas, such as Table Bay and Saldanha Bay, pollution levels around the South African coast are low by international standards (CSIR, 1979; Brown, 1987a). The chief concern to date is the crude oil pollution resulting from considerable tanker traffic around the Cape of Good Hope, as well as from accidents involving shipping of all kinds (CSIR, 1979; Brown, 1985a). Concern has also been expressed over sewage and storm-water contamination of beaches (Brown, 1987a,b; W.O.K. Grabow, personal communication); however, this concern revolves around possible health hazzards and there is no record of any detectable impact on the intertidal organisms themselves. In any event, the data which provide the material for the present review were all obtained from chemically unpolluted areas and particularly from beaches around Ou Skip, on the west coast, well away from industrial influences and high human population densities.

III. General Biology

A. *Introduction*

The genus *Donax* Linn., containing some 64 species, is the central and largest genus of the family Donacidae, which is placed within the superfamily Tellinacea. The animals are autobranch Bivalvia which are neither very primitive nor highly specialized. They are dominant burrowers on sandy beaches throughout the tropics and subtropics, with a few species, including *D. serra*, favouring temperate shores (Ansell, 1983). Nine species referred to the genus are recorded from southern Africa (Kilburn and Rippey, 1982). Our knowledge of the biology of *Donax* has been reviewed by Ansell (1983), while Donn (1986a) has edited a multi-author volume devoted to the genus in southern Africa.

The genus *Bullia* Gray (in Griffiths, 1833), originally erected to distinguish a group of species thought to be intermediate between *Buccinum* and *Terebra*, is a member of the stenoglossan family Nassariidae, which belongs to the most advanced superfamily of the neogastropods, the Buccinacea. It comprises about 25 species of small to medium-sized whelks, confined to sandy substrata in the intertidal zone and in shallow water. The possession of a broadly-expanded, thin foot is characteristic, as is the absence of eyes. Like *Donax*, most *Bullia* species occur in the tropics and subtropics, while a few, such as *B. digitalis*, are dominant on temperate sandy beaches in the southern hemisphere. The biology of the genus has been the subject of a review by Brown (1982a), subsequent papers dealing mainly with nutrition, respiratory physiology, pedal musculature and burrowing. The systematics of the group is reviewed by Cernohorsky (1984).

B. *Tidal Migrations*

It is typical of the macrofaunal species on exposed sandy beaches that they display tidal migrations up and down the slope, thus maximizing food resources and possibly avoiding predators (Brown, 1983). *B. digitalis* provides an example of this behaviour (Fig. 3), at least a proportion of the population migrating up and down the beach on every tide (Brown, 1971b, 1982a; McLachlan *et al.*, 1979b). The proportion actually migrating depends on the nutritional state of the animals, the availability of food (carrion) on the beach, beach slope, wave action and probably other factors as well. It may be as low as 12% (Brown, 1971b) or it may involve most of the population (McLachlan *et al.*, 1979b). Females carrying eggs do not migrate, however, but remain buried in the sand below low water mark (Brown, unpublished).

Intertidal members of the genus *Donax* are also typically migratory (Ansell, 1983) but *D. serra* is unusual in migrating only intermittently, chiefly on spring tides, with some migration at neaps, daily tidal changes being largely ignored by the adults (McLachlan and Hanekom, 1979; Donn *et al.*, 1986). Juveniles migrate more frequently than the adults, this behaviour decreasing with increasing size. The largest adults (shell-length 70–88 mm) appear not to migrate at all.

In both animals, emergence from the sand and expansion of the foot as an underwater sail is stimulated by water currents (Brown, 1961, 1971b, 1982a; Stenton-Dozey, 1989). The direction of tidal flow then determines whether the animal surfs up or down the beach. Reburial occurs when the water velocity is low enough to permit burrowing, although burrowing is frequently delayed in *Bullia* while the whelk crawls in search of food (Brown, 1971b). Burrowing also occurs when the animal is stranded on saturated sand but not on drier sand, which is more resistant to penetration.

Although there is no endogenous clock in either animal which corresponds to tidal periods (Brown, 1961, 1982a; Ansell, 1983), it seems probable that cues other than water currents may lead to emergence at appropriate times within the tidal cycle. This is particularly so as conditions near the top of the intertidal slope are very different from those towards the bottom. Trueman (1971) suggested that emergence from the sand on a rising tide may, in *Donax*, be a response to increasing liquefaction of the sand, a factor which possibly triggers the emergence of other psammophiles as well (Cubit, 1969). The emergence of *Donax* from their upper positions on the shore may be related to increasing periods of non-saturation between waves (Turner and Belding, 1957; Wade, 1967; Trueman, 1971). Branch and Branch (1981) have assumed that these changing factors are of importance in the migrations of *D. serra*, and they could also play a part in the migratory cycles of *Bullia*.

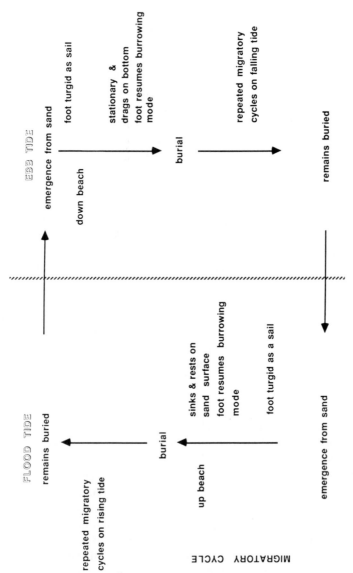

Fig. 3. The tidal migratory cycle of *Bullia digitalis* in response to ebb and flow.

There is no evidence of migrating *D. serra* or *B. digitalis* being left behind by the falling tide, an observation which attests to the efficiency of the mechanisms involved. In this *B. digitalis* differs from *B. rhodostoma*, which is frequently found buried in sand above the water (Brown, unpublished).

C. *Distribution on the Beach*

Tidal migration implies that the zonal distribution of the animals changes with the tidal cycle. The fact that in both species only a proportion of the population migrates, while some individuals only migrate partially and others not at all, means that the population becomes more widely dispersed as the tide rises and compressed seawards as the tide falls. The extreme upper region of the intertidal zone is, however, avoided.

The relative distribution of juveniles and adults is not the same on all beaches but varies according to conditions. In *D. serra* on the south coast, juveniles tend to occupy a position lower down the slope than the adults, at least while the tide is out, while on the Atlantic (west) coast the reverse is true (Bally, 1981, 1983b; Birkett and Cook, 1987). There has been much specula-tion as to the reasons for this size-related distributional difference. The two coasts present differences not only in microhabitat but also in food resources. Moreover, there is the possibility of competition with *D. sordidus* on the south coast but not on the west (Donn, 1986b). However, it may be significant that at the extreme northernmost end of its distribution, along the Skeleton Coast of Namibia, the zonation pattern of adults and juveniles again reverses, being the same as for the south coast. This fact has led Baily (1986) to postulate that the zonation is temperature related.

In *B. digitalis* also there tends to be some separation between adults and juveniles (McLachlan, 1980b; Brown, 1982a). On some beaches the juveniles occur somewhat lower down the beach than the adults during low water, catching up with them as the tide rises (McLachlan *et al.*, 1979b), while on other beaches there is often horizontal, long-shore separation (Brown, 1971b). There can be little doubt that these differences in distribution result from differences in wave action, water currents and beach slope while the animals are surfing, for they then have no control over direction or distance covered and are sorted virtually passively. This is also doubtless the reason for the virtual segregation of *B. digitalis* and *B. rhodostoma*, where both occur on the same beach, for the latter species is carried further in a water current than is the former (McGwynne, 1980; Brown, unpublished).

The longshore distribution of both *Donax* and *Bullia* is typically discontin-uous, areas of high population density alternating with areas in which no individuals of the species are to be found. Such discontinuity is, in fact, frequently observed among the macrofauna of exposed sandy beaches (Brown, 1983). While this may also be attributable in part to sorting by water currents, other factors may be relevant as well. Thus in the case of *Bullia*, groups are likely to be formed, or at least maintained, by feeding behaviour. A food item such as a stranded jellyfish draws the group together to feed, tending to prevent dispersion (Brown, 1982a). Another factor which may well be of considerable importance with regard to discontinuous distribution is the permeability of the sand. In order to establish themselves in a particular area of the beach, the animals must be able to dig into the sand to a certain critical extent before the arrival of the next wave. Where resistance to penetration is high, this may not be possible and the animals will then be swept away to burrow elsewhere. Stenton-Dozey (unpublished) has noted that where the sand is very compact no *D. serra* occur.

Donn (1986b, 1987) has studied temporal changes in distribution of *D. serra* in the eastern Cape and finds higher abundances on flatter slopes. He points out that these regions on cusp horns present more uniform wash and slower current speeds than adjacent areas in rip bays. He suggests that *D. serra* may be able actively to select for flatter slopes. Stenton-Dozey (unpublished) has also observed that adult *D. serra* tend to aggregate on the cusp horns during LWS on the west coast and postulates that food availability may be greatest in these areas. No such tendencies have as yet been noted in *Bullia*.

On some beaches, particularly on the west coast, the population of either *D. serra* or *B. digitalis*, or both, may be centred subtidally (Brown, 1971b; Birkett, 1986). Indeed *B. digitalis* is opportunistic to the extent that subtidal populations may be found off beaches whose slopes and wave activity normally preclude colonization; after storms have cut back the beach, reducing both slope and wave impact, the animals may be found intertidally (Brown, 1971b). *Donax*, being far less mobile, does not appear to indulge in this behaviour. It may, however, be noted that very small, juvenile *D. serra* are sometimes found on extremely high energy beaches, such as that at Noordhoek, where adults do not occur intertidally (Boland, personal communication). It could be that these juveniles are later transported to other beaches or that they eventually perish. However, it appears more likely that they originate from a population in the deeper subtidal zone, a population to which they return as they mature.

D. *Nutrition*

Although the two species differ completely in their methods of feeding, *Donax* being a filter feeder and *Bullia* primarily a scavenger (Ansell, 1981,

1983; Brown, 1961, 1971b), both may be described as opportunistic feeders with an extremely wide diet. In *Donax*, feeding takes place most commonly with the opening of the inhalent siphon flush with the surface of the sand. It may face shorewards or it may face the incoming waves, probably dependent on the physical properties of the water flow. The exhalent siphon may face the same direction as the inhalent or may face the opposite direction. On the south coast the diet of *D. serra* consists mainly of surf diatoms, and particularly *Anaulus*, while on the west coast it consists largely of detritus derived from phytoplankton. However, a wide variety of particles, with respect to both nature and size ($< 40 \mu$m), appears to be acceptable to the animal, including bacteria (Stenton-Dozey, 1989; Matthews *et al.*, 1989). It shows considerable variation in filtration rate, even in a single individual, and filters in pulses rather than continuously.

Bullia depends largely on carrion for its food but will accept any animal material, from siphonophores and other Cnidaria to mammals (Brown, 1964a, 1971b, 1982a). It will also turn predator, eating crustaceans and molluscs, including *Donax*, and possessing behavioural adaptations which allow it to prey efficiently on small amphipods and isopods (Gilchrist, 1916; Brown, 1971b). In addition, *B. digitalis* harvests an algal garden growing on its shell (da Silva and Brown, 1984; Harris *et al.*, 1986) and is also capable of absorbing dissolved organic matter from sea water, through the surface of its foot as well as through the pallial complex (Colclough and Brown, 1984). In fact *B. digitalis* presents an extreme example of the opportunism and variety of feeding methods that can be regarded as typical of macrofaunal species on exposed sandy beaches (Brown, 1983).

The nutritional problems faced by the two species are, however, different and must be solved in different ways. Food particles suitable for *Donax* are virtually always present in the water bathing the beach and represent a continuously renewable resource but both quality and quantity vary. *Donax* must thus be adapted to taking in the highest quality food, when it occurs, with the least effort. On a diet of detrital foam, assimilation efficiency seldom exceeds 50%, although it may reach 71% on a diet rich in organics such as cultured algae in the laboratory (Stenton-Dozey, 1989).

Bullia faces a different problem in that the supply of carrion to the beach is highly erratic and months may pass without such food being available (Brown, 1964a, 1982a; McGwynne, 1980). When carrion is washed up, however, it commonly arrives in quantities that greatly exceed what can be eaten by the population before it is pushed up the shore, out of reach, by the next high tide. The whelk must thus be adapted to survive long periods without carrion and to eat as much of it as possible during its passage through the habitat. In this regard, extremely rapid locomotory activity, involving not only crawling but also surfing (Brown, 1961, 1982a; Trueman and Brown, 1976) allows it to reach its food quickly and it has long been known that an adult will ingest up to a third of its own tissue weight in food during a single ten-minute meal in the field (Brown, 1961). This latter ability is made possible not only by efficient and powerful radula movements but also by the animal's ability to ingest at a fast rate by suction exerted by the proboscis (Trueman and Brown, 1987a). An important feature of the feeding behaviour of *Bullia* is that the whelks tend to anchor themselves to the food, so that they are not washed away from it by the waves and wash. If the food is relatively large and has a hard exterior (e.g. a bivalve), anchorage is achieved by suction produced by the sole of the foot. If the food is large but soft (e.g. a jellyfish), the proboscis itself provides anchorage by being thrust deep into the tissues. Small pieces of food, on the other hand, may be dragged below the surface of the sand so that they cannot escape (Stenton-Dozey, unpublished). The whelk also enjoys a high assimilation efficiency as compared with *Donax* (Stenton-Dozey and Brown, 1988).

The digestive carbohydrases of *Bullia* have been studied by Harris *et al.*, (1986), while Krohn (personal communication) has undertaken a prelimi-nary study of carbohydrases in the digestive system of *D. serra*. For both animals, α-amylase activity was found to be high, cellulase activity somewhat lower. Laminarinase activity was high in *Donax* but very much lower in *Bullia*. The generally higher values gained from *Donax*, for all three of the above classes of enzymes, are consistent with that animal's herbivorous diet, while the plant intake of *Bullia* merely supplements its carnivorous diet and appears to be confined to the cropping of its algal garden. In both species the digestion of carbohydrates appears to be aided by gut bacteria, although in *Bullia* only cellulase-producing bacteria were identified and in *Donax* only bacteria capable of digesting starch. These results require confirmation.

Rates of ingestion, assimilation efficiencies, energy budgets and ecological efficiencies are dealt with in detail in a later section of this review.

E. *Sensory Mechanisms*

Sensory mechanisms concerned with feeding are particularly well developed in *Bullia*. Brown and Noble (1960) demonstrated that the osphradium was the organ concerned with distance chemoreception in this animal; its ultrastructure was later investigated by Newell and Brown (1977). Brown (1971b) showed that the osphradium was particularly sensitive to volatile amines such as trimethylamine and studied the sequence of responses which lead to feeding. Hodgson and Brown (1985, 1987) identified the leading edge of the propodium as being important in contact chemoreception and showed that the application of amino acids to this region of the foot resulted in eversion of the proboscis, while the application of amines did not. Hodgson (in preparation) has discovered innervated ciliary cells in the leading edge of the propodium which may well be the relevant sensory receptors. The role of the subradular organ, which is present but diffuse and not well developed in *Bullia* (Brown, 1982a), is not yet clear and it is not possible to attribute any chemoreceptory function to it at this stage.

The process of acquiring food is less complex in *Donax* than in *Bullia*, as the former animal is far less mobile and does not crawl in search of food. There is no osphradium and no subradular organ but Hodgson and Fielden (1984) have shown that both inhalent and exhalent siphons bear ciliated tufts on their inner and outer surfaces. It is suggested that some of these ciliated cells are primary receptors (Hodgson and Fielden, 1984; Hodgson, 1986). Many of these, and particularly those on the outer siphonal surfaces, are likely to be tactile but as more than one type of ciliated cell is involved it is equally likely that some are chemosensory.

Neither species possesses eyes, this being typical of bivalves but more unusual in gastropods. *Donax* displays a typically molluscan "shadow response", withdrawing foot and siphons and closing the shell, if a shadow suddenly falls upon it (Trueman and Brown, 1985). It was thought for many years that *Bullia* was completely insensitive to light but Brown and Webb (1985) have shown that activity in non-buried *B. digitalis* increases temporarily in sudden total darkness after a period of bright illumination. While the shadow response of *Donax* may clearly be of some survival benefit in offering protection against predators, the significance of *Bullia*'s dark response is far from clear.

The tactile sense must be expected to be of importance in both *Bullia* and *Donax*, and the siphons of both animals are indeed very sensitive to touch. Both the dorsal and ventral surfaces of the foot of *Bullia* display numerous ciliated tufts (Hodgson and Cross, 1983). While those examined on the metapodium are not innervated and are thus not primary receptors, many of those on the propodium, and particularly on its leading edge, are innervated (Hodgson, personal communication). At least some of these could thus be tactile. The foot of *Donax* appears to be more sensitive to touch than that of *Bullia*. Ciliated tufts are present, particularly at its distal margin, and are being investigated by Hodgson (in preparation).

F. *Reproduction*

Donax and *Bullia* present a marked contrast as far as reproduction is concerned. In *Donax*, fertilization is external, as in all bivalves, and a free-swimming planktotrophic larva develops from the zygote, passing through trochophore and veliger stages before settling on the beach as spat. On the other hand, in *Bullia* fertilization is internal, copulation taking place either in the surf or within the sand (Brown, 1971b, 1982a), and the larval stages are suppressed, being passed inside the egg (da Silva and Brown, 1985), so that they hatch within the egg capsule as miniature, crawling whelks.

Donax serra produces very large numbers of gametes throughout the year and spat may be found on the beach at all times, although seasonal peaks have been identified (van der Horst, 1986). On the west coast, these peaks are often indistinct or absent (de Villiers, 1975a,b; Birkett and Cook, 1987). Sometimes peaks can be identified in early autumn and spring but there is no consistency from year to year. On the south coast, however, seasonal breeding peaks are much more pronounced (van der Horst, 1986).

Bullia digitalis, by contrast, has a very restricted breeding season and appears to produce only a single batch of eggs per year (Brown, 1982a). The developing eggs are either held in numerous small egg capsules attached loosely to the under surface of the maternal foot or are contained in a single large case within the sand and brooded over by the female (Brown, 1982a; da Silva and Brown, 1985). The batch may contain up to 40 000 eggs, all of which are potentially viable. It may be noted that this is a very large number of fertile eggs for a *Bullia* species and that *B. tenuis* produces only about 60 young a year, all but one of the eggs laid in each capsule being nurse eggs (Brown, 1985b).

In view of the large numbers of oocytes available per female, the fact that the oocytes are small and contain little yolk, that fertilization is external and that free-swimming larvae are produced, that there is an extended breeding season, that the sex ratio is 1:1 and that food resources are not really limited, van der Horst (1986) considers *D. serra* to be a typical r-strategist. *B. digitalis* differs from *Donax* in every one of the above characteristics (though not always to the extent that *B. tenuis* does) and is thus clearly a k-strategist.

There is a considerable amount of information on reproduction in both *Donax* and *Bullia* which has not yet been published and is regrettably not available to the authors of this review.

G. *Growth and Production*

Growth rates and productivity of *D. serra* have been assessed by McLachlan and Hanekom (1979) for beaches in the Eastern Cape, where the standing crop is up to 7000 g/m (McLachlan, 1977a,b). The resulting growth curve (Fig. 4) is very similar to that arrived at by de Villiers (1975b) for *D. serra* populations on the west coast. Although data for *B. digitalis* are not nearly as detailed, it is apparent that growth is very much slower, a shell length of

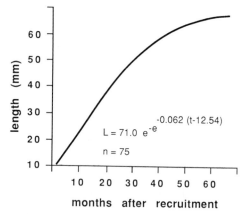

FIG. 4. Growth curve for the population of *Donax serra* on Sundays River Beach in the Eastern Cape Province (after Donn, 1986b).

60 mm only being attained after more than 20 years (Brown, 1982a; Kilburn and Rippey, 1982), as compared with about five years for *D. serra*. Indications are that the growth rate of *B. digitalis*, at least on the south coast, is very similar to that of *B. rhodostoma*, where the von Bertalanffy growth equation (for King's Beach in the Eastern Cape) is $L_t = 47 \ (1 - e^{0.19(t + 0.23)})$ (McLachlan *et al.*, 1979a). The hypothesis that the growth of *B. digitalis* is slower on the west coast than on the south remains to be tested.

Annual production of *D. serra* on Eastern Cape beaches has been determined by two methods, both of which yielded P/B ratios of 0.58 (McLachlan and Hanekom, 1979; Donn, 1986b). In contrast, the P/B ratio for *B. rhodostoma* (on Maitland River Beach) was only 0.06 (McLachlan *et al.*, 1979a). Similar P/B ratios are indicated for *B. digitalis* on the south coast, although the ratio may be lower on the west coast.

H. *Predation*

Donax and *Bullia* have a number of aquatic predators in common, among them the swimming crab *Ovalipes punctatus* (de Haan) (du Preez, 1984). The crab uses a number of methods to extract the flesh of these molluscs but is more successful with small to medium-sized animals than with large individuals. A number of fishes also move in with the tide to prey on *Donax* and *Bullia*. These include the teleosts *Lithognathus lithognathus*, *L. mormyrus*, *Pomadasys commersonni*, *Coracinus capensis* and *Rhabdosargus holubi*, as well as elasmobranch fishes such as *Rhinobatis* and the bull ray *Myliobatus aquila* (Brown, 1964a, 1971b, 1982a; McLachlan *et al.*, 1979a; Lasiak, 1983,

1984; du Preez, 1986). *Bullia* and small *Donax* are crushed by these fishes, whereas large *Donax* most commonly have their siphons bitten off. This does not normally result in the death of the mussels, and the siphons are rapidly regenerated (Hodgson, 1982, 1986). Du Preez (1986) has calculated that the crab *Ovalipes* can consume on average the equivalent of 2275 kJ/crab/y, while the fishes account for some 100 000 kJ/m/g(body weight)/y. *Bullia* must be more vulnerable to this predation than the more deeply-buried *Donax*, but this is offset by *Bullia*'s more consistent tidal migrations, the population tending to move up the shore as the tide rises, making it difficult for aquatic predators to reach them (Brown, 1983). While such migrations reduce the effects of predation by aquatic carnivores, they may be expected to increase vulnerability to terrestrial predators, particularly birds. However, in contrast to *Donax*, which is preyed upon by several species of the avifauna, *Bullia* appears not to be attacked by them (Brown, 1964a, 1971b, 1982a).

I. Environmental Tolerances

Although a number of environmental tolerance limits have been established for either *Bullia* or *Donax*, the only comparison that can be made between the two animals relates to temperature. Brown (1961) considered briefly the upper temperature tolerances of *B. digitalis* and *B. laevissima*, while McGwynne (1980, 1984) performed similar experiments on three species of *Bullia* from Eastern Cape beaches. Ansell and McLachlan (1980) carried out more sophisticated tests on *B. rhodostoma* and on *Donax*, and the effects of temperature on *Donax* have been studied in detail by Stenton-Dozey (1986, 1989). The latter investigations relate specifically to the population at Ou Skip, on the west coast. The animals display extreme sensitivity to temperatures above 31°C, the smaller mussels showing marginally better survival at acute temperature exposure than the adults. The incipient temperature (i.e. the highest temperature to which the animals can be exposed continuously for an indefinite period without increasing mortality rate) is in the region of 25°C for all size classes. The LT_{50} data presented by Ansell and McLachlan (1980) for south coast *D. serra* are directly comparable with above results; on the south coast, large individuals show a higher temperature tolerance than small individuals, in contrast to those on the west coast. The difference is clearly related to differences in tidal exposure. Temperature-related burrowing responses also differ between the two coasts. In both west and south coast populations, emergence from the sand is a common reaction to heat stress over a prolonged period. Finally, there is evidence that in the field the sand above the animal (up to 30 cm) may provide significant insulation against thermal shock—a particularly important phenomenon on the west coast, with its rapidly fluctuating temperature regime.

No such insulation is afforded to *Bullia*, which occurs at or just below the surface. It is thus not surprising to find that the intertidal species of *Bullia* are more tolerant of high temperatures than *Donax* on the same beaches (McGwynne, 1980; Ansell and McLachlan, 1980). Furthermore, where several species of *Bullia* occur on the same beach, their temperature tolerances are roughly correlated with tidal level (Brown, 1961; McGwynne, 1980, 1984). The latter author gives the "thermal death point" for *B. digitalis* as 33°C. Large individuals show a greater thermal tolerance than do small individuals. As in the case of *Donax*, emergence from the sand is observed in *Bullia* as a response to heat stress.

J. *Effects of Pollution*

The effects of pollution on several species of *Bullia*, including *B. digitalis*, have been studied quite extensively (summarized in Brown, 1982c). A common initial effect, at low concentrations of pollution, is a failure of the responses that lead to feeding. This lack of attraction to food is followed, at higher concentrations, by emergence from the sand or a refusal to burrow if already emerged (Brown, 1986a). The animal lies on its back, with the foot held fully expanded and turgid, as it is in surfing; the foot is waved from side to side, which encourages transport by waves and currents to possibly less polluted areas. At higher concentrations of the pollutant, the normally yellowish foot starts to turn grey and becomes creased, while foot waving declines. This state of "lethargy" eventually gives way to paralysis, which continues until the animal dies.

  Donax serra is inclined to react to sudden stress, including that brought about by acute chlorine pollution, by the immediate withdrawal of the foot and siphons and closure of the shell (Stenton-Dozey, 1989). This is, in fact, a common escape response among bivalve molluscs suddenly exposed to chemical pollution, as well as to drastic changes in salinity (Trueman, 1983). Subacute, prolonged stress leads to emergence from the sand, although this does not happen as soon or as consistently as it does in *Bullia*. Indeed a major difference between the two animals faced with unpleasant conditions is that the initial response of *Donax* is closure, while that of *Bullia* is emergence. Retraction into the shell is a very rare event in the latter species.

IV. Locomotion

A. *Introduction*

The ability to burrow rapidly and to emerge from the substratum is an essential prerequisite for macrofaunal life on sandy, wave-swept shores. *B. digitalis* and *D. serra* provide excellent examples of this ability, showing a number of similarities which arise by convergent evolution.

All soft-bodied animals burrow in an essentially similar manner, based on the alternate application of two types of anchorage, each of which allows another part of the body to move into the sand, so producing a stepping motion, each step being termed a digging cycle (Trueman, 1983). The first, or penetration, anchor is formed by the dilatation of part of the posterior region of the body and prevents the animal from being pushed out of the ground as it thrusts downwards. The second, or terminal, anchor is formed by the swelling of the anterior region so as to allow the posterior to be drawn down. To permit such changes in shape, soft-bodied burrowing animals have characteristically developed large, fluid-filled cavities that effect both changes in shape and the transfer of muscular force (Trueman, 1975). The Bivalvia are burrowing molluscs *par excellence*, for they are primitively infaunal, with laterally compressed shell and blade-like foot with which to thrust forward into sand or mud (Trueman, 1976; Runnegar and Pojeta, 1985) (Fig. 2). Gastropods, on the other hand, are primarily adapted to hard, rocky surfaces and secondary adaptation to an infaunal life has evolved separately in a number of taxa, for example Nassariidae, Naticidae and Conidae. Slowstepping, with alternate application of penetration and terminal anchors, as occurs in *Bullia*, is an adaptation of normal gastropod locomotion, in which numerous pedal waves of small amplitude occur, for an infaunal mode of life.

An important difference between the two molluscs is that *Bullia* moves immediately obliquely into the sand, at an angle of only 10–15° from the horizontal, and until the shell is just covered (Fig. 6), while *Donax* first extends the foot into the substratum with a stabbing or probing motion, which greatly reduces the resistance of the sand, until it has penetrated sufficiently to obtain a secure anchorage to pull the shell upright, the animal then burrowing vertically downwards (Fig. 5). The deeper burrowing of *Donax* entails motion into more resistant, closely packed sand than *Bullia* encounters nearer the surface.

To live in sand exposed to wave action, it is necessary that the animals are able to avoid being buried too deeply by sand deposition (Trueman and Ansell, 1969). Both these molluscs achieve this, *Bullia* by normal stepping motion in an upward direction, and *Donax* by reversal of direction by means of powerful downward thrusts of the foot.

Bullia crawls over the surface of wet sand using the same locomotory mechanisms as in burrowing and also moves rapidly up and down the beach by using its broadly-expanded, turgid foot as a sail in the surf (Brown, 1961; Trueman and Brown, 1976) (Fig. 1). *D. serra* also surfs. Mature individuals (with a shell length greater than 2.5 mm) probably emerge from the sand only

exh inh

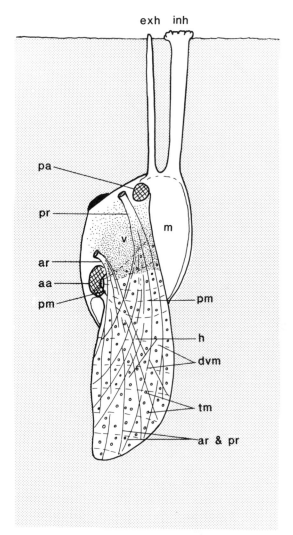

FIG. 5. Diagrammatic longitudinal section of *Donax serra* in sand, with foot extended (× 1.0). aa, anterior adductor; ar, anterior retractor muscle; dvm, dorso-ventral muscle; exh, exhalent siphon; h, haemocoel; inh, inhalent siphon; m, mantle cavity; pa, posterior adductor; pm, protractor muscle; pr, posterior retractor muscle; tm, transverse muscle; v, viscera.

once or possibly twice per lunar cycle but smaller *Donax* may surf more frequently (McLachlan and Young, 1982; Stenton-Dozey, 1989), in a manner similar to that of the tropical surf clam *D. denticulatus* (Trueman, 1971). This behaviour requires the ability not only to emerge from the sand but also to

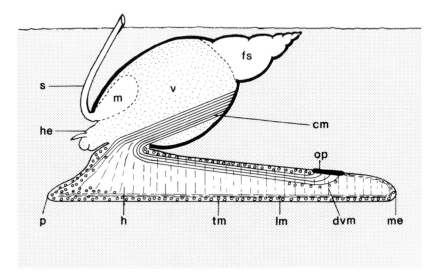

FIG. 6. Diagrammatic longitudinal section of *Bullia digitalis* in sand, with foot extended and siphon protruding as during locomotion (× 2). cm, columellar muscle; fs, free space at apex of shell, filled with water when foot extended; he, head; lm, longitudinal muscle; me, metapodium; op, operculum; s, siphon; other lettering as in Fig. 5.

burrow rapidly in a new location before being washed away by a following wave.

B. *Functional Anatomy*

In *Bullia* the disc-shaped foot is extremely flattened, with a clearly defined pedal sinus (Fig. 6) that functions as a hydrostatic skeleton. This is in contrast to most gastropods, including *Patella* (Jones and Trueman, 1970), *Busycon* (Voltzow, 1985) and *Nassarius* (Hodgson and Trueman, pers. comm.), where muscle fibres predominate in the foot and blood is restricted to the intramuscular space, the foot functioning as a muscular hydrostat (Kier, 1988).

In addition, the foot of *B. digitalis* contains small aquiferous spaces in the posterior region, which open directly to the ventral surface (Brown and Turner, 1962; Brown, 1982a). These aquiferous spaces are entirely separate from the haemocoelic cavity and resemble the much more extensive spaces of *Polinices* (Russell-Hunter and Russell-Hunter, 1968; Trueman, 1983).

The pedal sinus of *Bullia* is supplied with blood from the anterior aorta and extends into the cephalic sinus, so that during locomotion these two chambers function together as a single cavity (Brown *et al.*, 1985a; Trueman and Brown, 1987b). Blood may leave this cephalopedal sinus either by way of a large cephalopedal vein or by pallial collecting vessels that carry blood to a large pallial sinus system (Brown, 1964b) (Fig. 7(a)). The cephalopedal and visceral veins are effectively elongate sinuses forming a simple tube, so that an increase in pressure in the pedal sinus may cause blood to flow through these veins into the visceral sinus system (Brown, 1964b, 1982a). This occurs when the foot retracts into the shell. At protraction, the foot is pushed out of the shell by the columellar muscle acting as a muscular hydrostat (Trueman and Brown, 1976; Brown and Trueman, 1982a,b), while blood flows into the pedal sinus either from the visceral sinus or from the heart, or both. The visceral musculature is, in fact, poorly developed and unlikely to exert much pressure on the blood; however, it may be argued that little pressure is required once the pedal muscles are relaxed. In any case, the foot would appear to inflate far too rapidly for this to be due solely to blood pumped into it by the heart.

For the foot to function as a fluid skeleton during locomotion, the blood in the cephalopedal haemocoel must remain at constant volume. No valves have been observed that might restrict blood flow from the foot but a network of muscle fibres in the neck tissue between the foot and the visceral mass may restrict blood flow from the foot (Trueman and Brown, 1976). The pedal musculature of *Bullia* consists of the extrinsic columellar muscle, which principally serves the operculum but also has many fibres passing to the propodium, and intrinsic muscles arranged in three planes: transverse, dorsoventral and longitudinal (Fig. 6). Longitudinal muscle near the sole of the foot has not been observed in gastropods moving by retrograde steps over hard substrata but the others have a similar function in stepping (Trueman and Brown, 1987b). Da Silva and Hodgson (1987) have demonstrated that two different types of muscle fibre occur in the foot of *Bullia*, fibres with smooth characteristics, which presumably sustain tension, and striated fibres which contract more rapidly in propodial movements. The functional roles of these fibres have been further discussed by Trueman and Brown (1987b).

Donax serra has a blade-like foot which can be totally withdrawn between the valves or extended downwards for approximately the same distance as the length of the valves. The mantle cavity acts as a compensation chamber at pedal retraction, when it is almost entirely filled by the foot (Trueman and Brown, 1985). The foot contains a haemocoelic cavity, the pedal haemocoel, which persists during retraction and is greatly enlarged by blood flowing into it during extension. The rate of blood flow required to transfer the blood from the posterior storage system, principally within the mantle lobes, through the anterior aorta to the foot (Fig. 7(B)), is well within the capacity

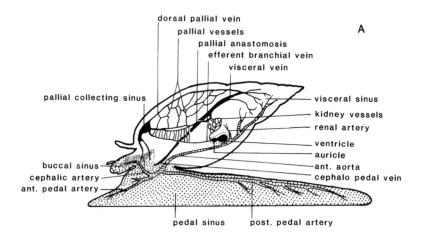

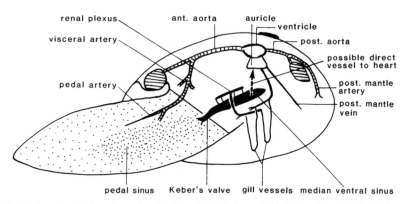

FIG. 7. Longitudinal sections showing blood circulation of (A) *Bullia digitalis* (after Brown, 1964b) and (B) *Donax serra* (after Trueman *et al.*, 1986). Arteries are indicated by hatched vessels, veins by solid vessels, while broken lines show possible alternative return pathways to the heart.

of the heart (Trueman and Brown, 1985; Trueman *et al.*, 1986). During protraction, a series of pressure pulses (of 300 Pa) occurs in the pedal sinus, each pulse corresponding to an extension of the foot by contraction of transverse muscles about a constant volume of blood retained in the foot by the action of Keber's valve. Thus in *Donax* pedal extension is achieved by the use of the pedal haemocoel in the manner of a classical hydrostatic skeleton, in contrast to *Bullia*, where the foot is pushed out of the shell by the columellar muscle functioning as a muscular hydrostat. During periods when the foot of *Donax* is inactive, blood returns to the heart with probably only an ebb and flow motion through the gills, as Knight and Knight (1986) have described for *Pholas*. However, at pedal retraction the rate of blood flow is greatly increased and the presence of a vessel from the median ventral sinus directly to the heart, bypassing the gills, is suggested (Fig. 7(B)) (Trueman *et al.*, 1986).

The pedal musculature of *Donax* may also be divided into two parts— extrinsic and intrinsic. The former, consisting of anterior and posterior retractors and the protractor muscles, are all inserted onto the valves and possess no trace of a muscular hydrostat system (Fig. 5). The retractors form a superficial network in the distal region of the foot and function to draw the shell into the sand; the protractor fibres pass superficially around the proximal part of the foot, where they aid pedal extension by constriction of this region and possibly also restrict blood flow from the foot. The intrinsic musculature consists of transverse and dorsoventral fibres (Figs 5 and 10), together with a variety of superficial fibres whose precise function is difficult to determine. Contraction of the powerful transverse musculature elongates the foot, particularly during probing movements, while the dorsoventral fibres are thought to prevent the blade-like foot from spreading dorsally and ventrally whenever high pressures are generated in the pedal haemocoel (Trueman and Brown, 1985).

The columellar muscle (in *Bullia*) and the retractor muscles (in *Donax*) both function to retract the foot into the shell and are homologous in this respect. The additional presence of a network of transverse fibres in the columellar muscle of *Bullia* enables this muscle to push the foot out of the shell, in contrast to the role of the fluid skeleton of the foot of *Donax*. In the feet of both genera, muscles in one plane (transverse in *Bullia*, dorsoventral in *Donax*) serve to maintain one pedal dimension constant while muscular antagonism in the other two planes effects stepping movements (Trueman and Brown, 1985, 1987b).

C. *Locomotory Behaviour*

The locomotory behaviour of both *Donax* and *Bullia* has been described by Trueman and Brown (1976, 1985). The essential movements are summarized in Fig. 8 and illustrated in Figs. 9 and 10. Both species represent good examples of burrowing animals, *B. digitalis* of 45 mm and *D. serra* of 65 mm shell length being able to penetrate completely beneath water-saturated sand in about 45 and 20 seconds respectively. Animals of these standard sizes have been used to estimate burrowing performance (Tables 1–3). Both animals burrow in a stepwise manner, the shell being static as the foot extends, followed by anchorage of the foot while the shell is drawn forwards. *Bullia* crawls over wet sand in the same stepwise manner. This consists of propodial extension, in a movement resembling a swimmer's breast stroke, and anchorage posteriorly (penetration anchor) by the expanded metapodium (when buried, also by the shell), followed by propodial dilation to form an anterior (terminal) anchor which allows the metapodium and shell to be drawn forwards (Fig. 9). Penetration of the sand is effected obliquely and only to sufficient depth for the short siphon to remain in contact with the overlying water. The events described constitute a single step or digging cycle.

   In *Donax*, extension and probing of the foot into the sand occur until sufficient anchorage is obtained for the shell to be pulled erect when the shell is progressively drawn downwards in a series of digging cycles. Each cycle (Fig. 8) involves anchorage (penetration) of the shell, while the foot thrusts

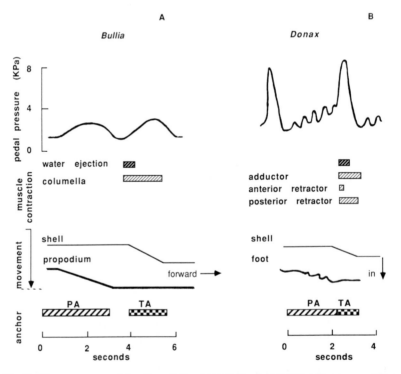

FIG. 8. Diagrams summarizing the activity of (A) *Bullia digitalis* (after Trueman and Brown, 1976) and (B) *Donax serra* (after Trueman and Brown, 1985) during a digging cycle, showing the relationships between pressure, muscular activity, movement and anchorage. PA, penetration anchor; TA, terminal anchor.

A   B

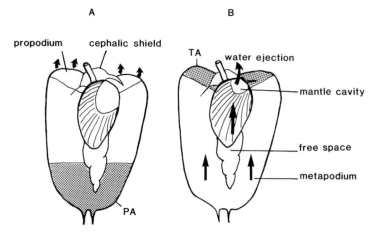

FIG. 9. Diagrams of *Bullia digitalis* showing movement during burrowing or crawling. (A) Extension of propodium in "breast stroke" (arrows), with cephalic shield pressing downwards and posterior of foot anchored (stipple). (B) The shell and metapodium drawn forwards (arrows), mantle cavity reduced in volume, with water being ejected forwards and the propodium anchored (stipple). Lettering as in previous diagrams. (After Trueman and Brown, 1976.)

downwards by contraction of the transverse muscles in antagonism to the distal retractor muscle fibres. Dilation of the foot to form a terminal anchor allows the shell to be pulled into the burrow. Anterior retraction is followed by contraction of the posterior retractors so as to impart a slight rocking motion to the shell. This is, however, by no means as marked as it is in more tumid bivalves (Trueman, 1983; Stanley, 1987). Blood is transferred to the distal part of the foot, effecting dilation, by adduction of the valves, which simultaneously causes water to be ejected from the mantle cavity, liquefying the sand adjacent to the shell (Fig. 10(B)) and thus facilitating its downward movement (C). Ejection of water from the mantle cavity is a common feature of all burrowing bivalves and a convergent adaptation is observed in *Bullia* (Figs. 8 and 9). Brown (1964b) observed that contraction of the columellar muscle to bring about pedal retraction was accompanied by jets of water, two-thirds of which was derived from the "free space", a chamber between the viscera and the apex of the shell (Fig. 6). Re-emergence of the whelk only occurs in water, for the animal refuses to draw air into its free space or into the small aquiferous spaces within the foot.

The function of these aquiferous spaces in *Bullia* has never been properly investigated. They are probably derived from pedal mucous glands by enlargement and elaboration (Brown, 1982a) and open to the ventral surface by a series of open channels. As these channels are not equipped with

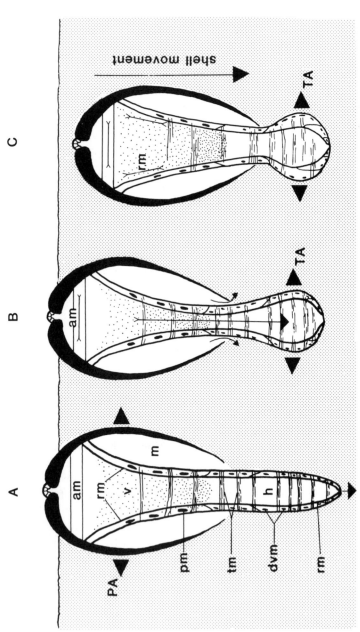

Fig. 10. Sections of *Donax serra* to illustrate the principal movements in burrowing. (A) Valves pressed outwards against the sand by the opening thrust of the ligament, to provide a penetration anchor while the foot probes downwards (arrow). The adductor muscle is relaxed and the foot extends and retracts by muscle antagonism distally about the pedal haemocoel. (B) Valves partially adduct (arrows indicate tension in the adductors), causing blood flow and dilation of the foot distally, forming a terminal anchor and water to be ejected from the mantle cavity to facilitate the motion of the shell through the sand in C. (C) Contraction of the retractor muscles (arrows rm) draws the valves down towards the anchored foot. Lettering as in previous figures.

sphincters, valves or any other structures which might impede flow (Brown and Turner, 1962), it is clear that water must enter and leave the aquiferous spaces in response to changing patterns of contraction and relaxation in the pedal musculature. It is thus possible that they serve to lubricate the ventral surface of the metapodium as this region of the foot is pulled forward during crawling and burrowing. There are no corresponding spaces in the foot of *Donax*.

In both *Donax* and *Bullia*, a standing pressure is maintained in the foot during burrowing and pulses are superimposed upon this during pedal extension and the forward movement of the shell. The amplitude of these pulses is always greater in *Donax* than in *Bullia* (Fig. 8). During pedal extension, the greater the pressure the more powerful the forward thrust, provided the penetration anchor is strong enough to prevent the animal from pushing itself out of the sand. The higher amplitude of the probing pulses in *Donax* suggests that anchorage by means of the gaping valves is the more secure method. A similar situation pertains in terminal anchorage, where pressure pulses in *D. serra* are about three times those observed in *B. digitalis*, indicating that more force is used by *Donax* in drawing the shell into the sand. It should be noted that *Bullia* burrows obliquely and shallowly into the sand, thus encountering much less resistance from the substratum than does the vertically, more deeply burrowing *Donax*. Some indication of this difference in resistance has been obtained by Brown and Trueman (unpublished), who measured the force necessary to penetrate the sand at different angles and depths.

An alternative means of transport, using the foot as a sail, has been particularly well developed in *Bullia* spp. (Brown, 1961, 1982a). In response to water currents, the surfing species of *Bullia*, including *B. digitalis*, emerge from the sand with turgid foot fully extended. Maximal surface area is achieved by the foot being made as thin as possible by contraction of the dorsoventral pedal muscles and the relaxation of the transverse and longitudinal fibres. Behavioural modifications allow emergence from the sand for tidal migrations up or down the beach and reburial when stranded (Brown, 1961; Ansell and Trevallion, 1969). *D. serra* is also able to migrate up and down the beach (Branch and Branch, 1981; McLachlan and Young, 1982), with the foot and siphons extended in a manner similar to that described for tropical species of *Donax* (Trueman, 1971). The extended foot and siphons increase the drag of the animal in the surf and enable burrowing to commence immediately a stable position is reached on the substratum. When the shell of the clam is longer than about 25 mm, it normally remains in the sand, adjusting its position in response to wave activity; however *D. serra* of all sizes may respond to stress by emerging from the sand and being moved passively by the waves (Stenton-Dozey, 1989).

McLachlan and Young (1982) give useful tables of burrowing performance in both *Bullia* (*B. rhodostoma* and *B. digitalis*) and *Donax* (*D. serra* and
D. sordidus) of various sizes and at different temperatures. They also discuss
the effects of wave height and frequency on migratory abilities. Burial on a
beach over which waves pass at about 10 second intervals is probably
restricted to small individuals, for these are the only animals to bury
themselves quickly enough to avoid being washed out of the sand by the
succeeding wave. Larger animals, which take longer to bury themselves,
would be restricted to calmer water beyond the surf and to the water-
saturated sand of the wash zone. The period of washes crossing 50% of this
zone ranges from 19 to 54 seconds on Maitland River Beach, in the Eastern
Cape, and once an animal is two-thirds buried, washes do not dislodge it.
Thus burial in the wash zone of this beach is no problem for either *B. digitalis*
or *D. serra* of the standard sizes cited in Table 1. *D. serra* of 25 mm have been
observed to surf up and down the beach between spring and neap tides
(McLachlan *et al.*, 1979b, Stenton-Dozey, 1989). The region of the beach
where they can burrow is limited by wave action in a manner similar to that
observed in respect of the tidal migrations of the tropical clam *D. denticulatus*
(Trueman, 1971).

D. *Energetics of Locomotion*

Quantitative comparisons between the locomotion of *B. digitalis* and *D. serra*
may be based on either respiratory studies or the estimation of mechanical
energy required for the burial of unit body mass. With many animals it is
difficult to achieve satisfactory results for burial in a respirometer chamber
but good results have been obtained for *Bullia* (Brown, 1979a,b, 1981).
Account must also be taken of anaerobic metabolism in estimating energy
cost.

The mechanical energy required for burial has been determined for species
of *Donax* (Ansell and Trueman, 1973) and *Bullia* (Trueman and Brown,
1976). The power required for burial is DU, where D represents drag and U
the rate of movement. The latter can readily be determined from film or
recordings, whereas drag can be ascertained by measuring the maximum
force exerted, using a force transducer attached by a thread to the shell. This
force must normally overcome drag and thus represents the maximum drag
encountered by the animal during burial.

The burrowing performances of *B. digitalis* and *D. serra* are compared in
Table 1, in which maximum effort has been determined at the midpoint of
burial of animals from Ou Skip, on the west coast of the Cape. The digging
period has been measured as the interval between entry of the foot into the

TABLE 1. BURROWING CHARACTERISTICS AND MECHANICAL ENERGY REQUIREMENTS

	Bullia digitalis	Donax serra
Shell length (cm)	4.5	6.5
Dry weight (g)	0.75	2.5
Duration of digging period (s)	45	20
Number of digging cycles	10	15
Mean rate of burial, U (cm/s)	0.10	0.325
Haemocoelic pressure (kPa)		
Maxima during a, pedal extension	2.0	2.0
b, shell retraction	3.0	8.0
Force of retraction, D (Newtons)	0.17	3.4
Power required for burial, DU $(J/s^2 \times 10^{-2})$	0.017	1.1
Mechanical energy for burial $(J \times 10^{-2})$	0.76	22
Mean mechanical energy for single digging cycle $(J \times 10^{-2})$	0.076	1.47
Weight specific energy for burial (J/kg)	10	88
Mechanical locomotory cost (J/kg/m)	222	1354

Data from Trueman and Brown, 1976, 1985 and from unpublished observations on *Bullia digitalis* of standard size (0.75 g).

sand (*Bullia*) or from the shell being drawn erect (*Donax*), until the shell just disappears beneath the sand surface. The much greater power requirement for burial, or mechanical locomotory cost, in *Donax* is apparent and may be accounted for in part by the different modes of burial in the two animals, *Bullia* moving obliquely, at an angle of only some 15° to the horizontal and to a depth of not more than 2 cm (Fig. 6), *Donax* moving vertically downwards and to a far greater depth (Fig. 5). Even on wave-washed beaches, the sand is more compacted with increasing depth, making deep burial relatively much more difficult, while in experimental laboratory conditions this difference may be even greater (Trueman *et al.*, 1966).

The lower haemocoelic pressure developed in *Bullia* at shell retraction and the lower retraction force developed should be noted (Table 1). These two factors are related, the stronger anchorage obtained in the deeper, more consolidated, sand being coupled with higher pedal pressures in *Donax*, which allow the shell to be drawn down with greater force but at a higher locomotory cost. The mechanical locomotory cost of burial in *D. serra* is comparable with that of other animals which burrow downward into sand, for example *D. incarnatus* (Ansell and Trueman, 1973; Trueman, 1983) and the polychaete *Nereis diversicolor* (Trevor, 1978); by contrast the energy requirement for shallow, oblique burrowing in *Bullia* is much less, as is the case also for the mole crab *Emerita portoricensis* (Trueman, 1983). There are few appropriate data on the cost of snail locomotion with which to compare *Bullia*, apart from estimates of the cost of crawling in terrestrial slugs, of about 900 J/kg/m (Denny, 1980). This may be taken as representative for snails moving by adhesive multiple pedal waves, for much of the energy is expended in the production of mucus. The value for slugs is much higher than for *Bullia* either burrowing (Table 1) or crawling over the surface of the sand, at a cost of about half that of burrowing (Brown, 1981, 1982b). *Bullia* crawls in a series of steps, each involving the whole foot, adhesion occurring by the use of the propodium and metapodium alternately (Trueman and Brown, 1987b). This raises the question of whether the "galloping" mode of locomotion in *Helix*, where only limited parts of the foot are in contact with the substratum (Trueman and Jones, 1977), with consequently less demand for mucus, is more economic of energy than is the use of characteristic direct or retrograde pedal waves (Trueman, 1983).

Comparisons have been made between estimates of locomotory cost based on respiratory studies and those from determinations of mechanical energy. Such comparisons present difficulties in that they assume metabolism to be entirely aerobic and, secondly, the conversion factor from chemical to mechanical energy must be established. In fact, molluscan metabolism is probably never 100% aerobic, although there is considerable evidence that in *Bullia*, metabolism is essentially aerobic even when the whelk is most active (Brown, 1984a; da Silva and Hodgson, 1987). This is not the case in *D. serra* and recent investigations have shown that pedal respiration is likely to be largely anaerobic (Trueman and Brown, 1985, 1987c). (A full discussion of respiration in *Bullia* and *Donax* appears in Section V.) Thus oxygen consumption in *Bullia* is likely to relate to locomotory effort, while that in *Donax* does not.

The mechanical energy cost of burrowing in *B. digitalis* has been determined by Trueman and Brown (1976), who also derived values for the chemical energy requirement based on the assumption that overall efficiency of locomotory activity (i.e. chemical-mechanical conversion and the mechanical efficiency of use of the foot) was about 20%. This value had previously been assumed in work on burrowing, for example by Ansell and Trueman (1973) as well as by workers on vertebrate muscle (Hill, 1939; Tucker, 1975). Brown (1979a) caused *B. digitalis* to burrow while the shell was attached by a thread to a force transducer, so measuring the retractive force exerted during each digging cycle. Oxygen consumption was measured simultaneously. Based on the probability of pedal metabolism during burial being almost entirely aerobic, he concluded that an overall, or gross, efficiency factor of only 6% was appropriate to the burrowing process in respect of the conversion of chemical energy to mechanical power. This value is used in the present article.

While on this basis the mechanical energy used by *Bullia* is comparable to

TABLE 2. THE COST OF BURROWING OF ANIMALS OF THE SAME CHARACTERISTICS AS IN TABLE 1

	Bullia digitalis	Donax serra
Mechanical energy for burial ($J \times 10^{-2}$)	0.76	22
Equivalent chemical energy at 6% conversion factor ($J \times 10^{-2}$)	12.6‡	365
Chemical energy for burial (determined by respirometry) ($J \times 10^{-2}$)	8.82‡	—
Routine (inactive) metabolic rate (J/kg/s)	2.9†	1.19*
Routine metabolism for duration of burial ($J \times 10^{-2}$)	10.4	5.9
Energy for burial (equivalent chemical energy): Routine metabolism for the same period	1.2:1	61.8:1

Data from ‡Brown, 1979a, †Brown, 1982b and *Stenton-Dozey, 1989.

the chemical energy requirement, attempts to estimate the cost of burial of *Donax denticulatus* by respirometry give values for oxygen consumption which, when converted using a 6% factor, were only 1/50th of the mechanical locomotory cost (Trueman, unpublished). Accordingly, we have not attempted to determine the energy required for burial in *D. serra* by respirometry and have assumed that the same relationship between chemical and mechanical energy obtains in both molluscs. It is useful to compare the mechanical requirement for burial with the routine metabolic rate in both species (Table 2). The inactive rate for *D. serra* was determined using unfed animals, buried and at rest in sand, with siphons extended and respiratory pumping taking place (Stenton-Dozey, 1989). Whereas in *Bullia* burial makes a demand which is slightly less than the metabolic rate, in *D. serra* the demand is about 60 times greater. Anaerobiosis during the measurement of routine metabolic rate may in part account for this higher value but, even if inactive metabolism was an order of magnitude greater, the difference between *Bullia* and *Donax* would still emphasize the high cost of rapid burial perpendicular to the sand surface. P. F. Newell (1970) and R. C. Newell (1979) have both observed that animals in their active state commonly use oxygen at twice the rate of those at rest. This condition is approached in *Bullia* and would be greatly exceeded in *Donax* were it not for the use of anaerobiosis.

E. *Time-energy Budgets*

Activity budgets may be drawn up following estimation of the cost of locomotion in *Bullia* and *Donax* (Tables 1, 2) and a knowledge of the behavioural activity of these animals on the beach. Brown (1981, 1982b) attempted to follow individual *Bullia* from their emergence on a falling tide to their final burial on the rising tide. He was successful with 18 large female animals, recording their activities (other than feeding) over a full tidal cycle. The oxygen consumption of similar whelks from the same locality was measured under control conditions to ascertain the energy costs of surfing, crawling, burrowing, emerging from the sand and remaining buried. Table 3a displays this data, the oxygen used, for example in burial, being added to the routine rate to give the total oxygen consumption for each event. The buried state is referred to in two categories—buried (observed) refers to periods between crawling and surfing, while buried (residual) indicates the unobserved part of the tidal cycle while the whelks were immersed. It is here that the greatest errors are liable to occur and Brown (1982a,b) has added two digging cycles per minute to allow for repeated adjustment of position in the sand in response to wave crash.

The comparable Table (3b) for *D. serra* differs from that of *Bullia* as only young individuals undertake tidal migrations with any frequency. The routine metabolic rate, representing a low level of water pumping, was determined in sea water without food; the animal adopts a higher metabolic rate as food becomes available (see section on Respiration). Just as in estimates for *Bullia* when buried, there is a large margin for error, as little is known about the clam's muscular activity when within the sand. In laboratory aquaria, *Donax* remains fairly static in the sand, with foot and siphons extended but it would seem reasonable to add the energy cost of two digging cycles per minute for adjustment of position in the field, as has been done for *B. digitalis*. This would represent an additional cost of about 357 J over a tidal cycle of 744 min (Table 3b). A small additional cost may be added to the total energy requirement for a tidal cycle, to represent the cost of emergence of the clams from the sand, being swept about by the waves to reburrow at another locality, which occurs possibly only once in each lunar cycle. Based on the cost of burial (Table 1) and assuming that the cost of emergence is equal to that of burial, this amounts to only about 7.3 J each 14 days, an addition to the tidal cycle cost of 0.26 J, which is thus not a major consideration. However, neither the duration nor the energy cost of surfing is known for *D. serra*. By analogy with *Bullia*, the energy cost of surfing with the foot extended and turgid is comparable with that of burial. Thus if the clam surfs for 10 min each 14 days, this is equivalent to a cost of some 110 J, or about 4 J per tidal cycle, which is again small in relation to the total energetic equivalent of 520 J per tidal cycle (Table 3). This is in marked contrast to the situation in *B. digitalis*, where the most active state may occur in migratory movements as the tide rises and falls. An individual of *Bullia* remaining buried during the tidal cycle expends about 108 J, while the most

TABLE 3. MEAN ENERGY EXPENDITURE OF STANDARD ANIMALS OF (a) *BULLIA DIGITALIS* AND (b) *DONAX SERRA* DURING SINGLE TIDAL CYCLES USING DATA DERIVED FROM TABLES 1 AND 2

(a) Activity	O_2 uptake (μg/min)	Duration of activity (min)	O_2 uptake for period (μg)	Energetic equivalent (J)
Transport in surf	20.8	14.5	302	4.31
Crawling	11.3	23.0	206	3.69
Burrowing	18.8	7.5	141	2.01
Emerging	18.8	3.8	72	1.05
Buried (observed)	9.3	42.0	391	5.57
Buried (residual)*	10.6	653.2	6924	98.60
Totals for tidal cycle		744	8090	115.23
For routine metabolism during single tidal cycle				98

Data from Brown (1982b). *Allows 2 digging cycles/min for adjustment of position in sand. See text.

(b) Activity	O_2 uptake (μg/min)	Duration of activity (min)	Energetic equivalent (J)
Routine metabolism	15	744	159
Buried†		744	357
Emergence, migration and reburial†		0.35	4
Total for tidal cycle			520

†Data derived from mechanical cost of burial; 2 digging cycles/min allowed for adjustment of position in sand. See text.

active animal uses some 133 J in the same period (Brown, 1982b). This is a low cost for being active during the tidal period, but it may be noted that about 12% of the cost of remaining buried is related to stepping to adjust position in the sand, a necessity on a high energy shore. The cost of locomotory activity during each tidal cycle represents an increase of some 17% over routine metabolism. In *Donax*, the cost of digging cycles, at a rate of two per minute, to adjust position exceeds routine metabolism by about 100%; however, routine metabolism has been estimated from oxygen consumption alone and we do not know the extent of anaerobiosis.

When the energetic equivalents for a single tidal cycle (Table 3) are expressed in terms of dry mass of the animals, the cost is 206 J/g for *Donax* and 153 J/g for *Bullia*. Yet the mechanical locomotory cost (Table 1) is an order of magnitude greater in *Donax* than in *Bullia*. This could be explained as indicative of a much higher activity level in *Bullia*, albeit at a much lower locomotory cost than in *Donax*. The latter cost is at least partly due to the shallower burrowing habit of *Bullia*, while the higher activity pattern of *Bullia* is associated with the erratic food supply and the need to exploit food resources as they occur. By contrast an adult *D. serra* has only to maintain its position on the shore to retain its potential as a filter feeder.

V. Respiration

A. *Gills and Respiratory Surfaces*

Bullia digitalis displays the characteristic neogastropod gill pattern, namely a single monopectinate ctenidium consisting of a series of filaments. A short siphonal canal, which extends to above the sand surface when the animal is buried, directs an inhalent current over the osphradium and the gill filaments, lying across the left side of the mantle cavity (Fig. 11). The exhalent current leaves the mantle cavity on the right side of the animal.

In *Donax serra* the mantle cavity is situated laterally on either side of the foot and viscera, in characteristic bivalve fashion, and water is led to and from the posterior of this cavity by long inhalent and exhalent siphons which may extend up to 7 cm in a buried, adult animal, thus obtaining access to well-oxygenated water above the sand. A pair of demibranch gills is situated in the mantle cavity on either side of the foot (Ansell, 1981) and serves for both respiration and feeding.

Yonge (1947) found that the ratio of gill area to wet body weight remained fairly constant, at $7-9 \, cm^2/g$, in all molluscs other than Bivalvia, where elongation of the gill filaments for feeding gave a ratio of 13.5 in *Mytilus edulis*. There is also a low rate of oxygen extraction from the water currents of only 5–9% in several bivalves (Hazelhoff, 1938) and Yonge thus suggested that the principal function of water flow over the gills was feeding rather than respiration. A recent investigation of gill size in prosobranchs (Trueman and Brown, unpublished) has shown a ratio of $5-7 \, cm^2/g$ in animals with paired bipectinate ctenidia (Pleurotomariacea), single bipectinate ctenidia (Trochacea) and single monopectinate ctenidia (Buccinacea). However, a strikingly low value of 1–2 was measured in species of *Bullia*. This could be accounted for by the use of the large disc-like foot as an auxiliary respiratory organ, Brown (1984a) having demonstrated that sufficient oxygen diffuses through its surface for all locomotory functions.

Surprisingly, in *D. serra* the ratio of gill area to wet body weight is no greater than in most of the prosobranchs examined, although it is somewhat higher than that of *Bullia* (Stenton-Dozey, unpublished). This is in marked

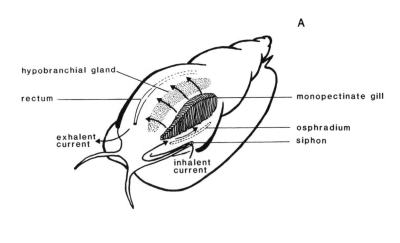

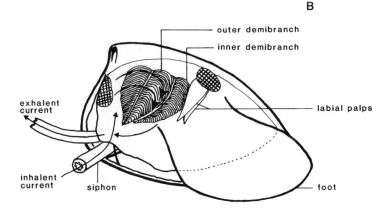

Fig. 11. Gill structure and flow of respiratory current in (A) *Bullia digitalis* (× 2) and (B) *Donax serra* (natural size).

contrast to the situation in *Mytilus* and may largely be accounted for by the retention of the large, muscular foot, which adds considerably to the body weight of the animal as compared with *Mytilus*. It has not been possible to demonstrate significant oxygen diffusion across the surface of the foot in *Donax* and oxygen tensions within it are low (Trueman and Brown, 1987c). The functions of the gills remain primarily the filtering of food, directing it towards the labial palps and rejecting extraneous particles via the exhalent siphon (Fig. 11(B)). It may be noted that, although the gill of *Bullia* is not a feeding organ, it too guides extraneous particles bound in mucus from the hypobranchial gland to the exterior.

No oxygen-carrying pigment can normally be detected in the blood of either animal. However, the blood of *Bullia* may acquire a blue colour, due to the presence of haemocyanin, after a period of exposure to certain forms of organic pollution (Brown *et al.*, 1985b). No such change has been observed in *D. serra*.

B. *Body Size*

While gaseous exchange is limited by the area of the respiratory surfaces, the volume of metabolizing protoplasm is directly proportional to body weight. The rate of oxygen uptake may thus theoretically be expected to be raised to the power of 0.67 of the body weight (Hughes, 1986). For *B. digitalis* this exponent (*b*) in the allometric equation $Y = aW^b$, relating oxygen consumption (*Y*) to body weight (*W*), has been arrived at for two widely separated populations. Adult females from Ou Skip (on the west coast) gave a value for *b* of approximately 0.6 (0.598) (Brown *et al.*, 1978) and this was found not to vary significantly between summer and winter (Brown and da Silva, 1979). *B. digitalis* from Algoa Bay, on the other hand, displayed lower values for *b*, ranging from 0.36 in summer to 0.55 in winter (Dye and McGwynne, 1980). It should be noted that the methods used in these two series of independent experiments were very different. Whelks from the Ou Skip population were suspended individually in a sealed chamber with the foot expanded and waving from side to side in a constant current generated by a magnetic stirrer set at 550 r.p.m. (Brown and da Silva, 1979), while the Algoa Bay animals were placed in unsealed chambers under liquid paraffin, with no attempt to achieve a constant level of activity (Dye and McGwynne, 1980). Nevertheless, differences between the two populations might be expected in view of the very different temperature regimes to which they are subjected (see p. 184). In particular, seasonal acclimation is clearly more appropriate on the south coast than on the west, with its rapid and extreme temperature fluctuations in summer.

The reason for the deviation of the *b* value from the "surface law" is not clear. Brown (1982a) suspects that in the field, under real conditions, the value may be still lower, for young individuals expend more energy per tidal cycle than do the adults, possibly resulting in a more gentle slope to the regression of weight against energy expenditure. However no data are available for values of *b* at different levels of activity, although it would seem that for the Ou Skip population the value should lie between 0.5 and 0.6 at all activity levels. The measured value of 0.598 has been assumed for all the activities listed in Table 4.

The weight exponent (*b*) of *D. serra* is closer to the predicted value, averaging 0.697 at ambient temperature when the animal is well fed and pumping water but reaching 0.759 during feeding (Stenton-Dozey, 1989).

TABLE 4. VARIATIONS IN THE RATE OF OXYGEN CONSUMPTION IN RELATION TO LEVEL OF ACTIVITY AS ANALYSED BY THE ALLOMETRIC RELATIONSHIP BETWEEN BODY WEIGHT AND METABOLISM FOR *BULLIA DIGITALIS* AND *DONAX SERRA* FROM OU SKIP BEACH, AT 15°C. IN THE ALLOMETRIC EQUATION, $Y = aW^b$, $Y =$ OXYGEN CONSUMPTION (ML O$_2$/H), $a =$ Y-INTERCEPT AS ML O$_2$/H/G ANIMAL, $W =$ DRY BODY WEIGHT (G) AND $b =$ WEIGHT EXPONENT

| Level of activity | *Bullia digitalis* | | | *Donax serra* | | |
	Description of activity	Regression a	b	Description of activity	Regression a	b
Standard	buried, ventilating	0.469	0.598	buried, ventilating	0.274	0.686
Routine	crawling, lab. current[1]	0.564	0.598	standard + digestion	0.373	0.697
Active	burrowing, surfing[2]	0.936	0.598	standard + digestion, feeding	0.454	0.759
Free existence	mean of above over 12.4 h tidal cycle	0.536	0.598			

[1] Foot expanded and waving in a constant current.
[2] Foot expanded and waving vigorously in a varying current.

The latter value approximates 0.75, the exponent arrived at by gross comparison of organisms ranging from Protozoa, through the invertebrates, to the largest mammals (Hemmingsen, 1960).

The fact that the weight exponents for both species are less than unity signifies that metabolic rate becomes relatively slower with increasing size. It is apparent from the weight specific metabolic rate, where $Y = aW^{b-1}$, that the difference in oxygen consumption with size is more marked in *B. digitalis* than in *D. serra*, the exponents being -0.402 and -0.303 respectively.

C. *Level of Activity*

In reviews of molluscan metabolic rates (Bayne, 1976; Newell, 1979; Bayne and Newell, 1983), much attention has been given to clarifying the relationship between oxygen consumption and levels of activity such as ventilating, feeding, burrowing, crawling and swimming. It is convenient, and generally agreed, to distinguish three activity levels—standard, routine and active. The routine rate is the least well defined and in fact covers a wide range of activities between standard (minimum) and active (maximum). Thus in comparing metabolic expenditure in *B. digitalis*, a very mobile carnivorous scavenger, and *D. serra*, a semi-sessile suspension feeder, different levels of activity need to be carefully defined. Standard activity is recognized in starved, immobile *Bullia* and *Donax* that are buried but ventilating.

Brown (1979a, b, c) has studied rates of oxygen uptake in *B. digitalis* at different levels of activity at 15°C, while Dye and McGwynne (1980) considered the oxygen consumption of relatively immobile (but not buried) individuals from a different population. The levels of activity tested related in particular to surfing, which approaches the active rate, burrowing and crawling. Feeding, or the presence of food, was found not to result in the greatly elevated rate of oxygen uptake measured in some whelks (Crisp *et al.*, 1978). While it was possible to arrive at an estimate for standard rate of oxygen consumption, no measurement of basal metabolic rate was possible as whelks buried beneath the sand in stagnant water showed a decrease in oxygen uptake with time, extremely low values being eventually reached (Brown, 1979c).

In *D. serra*, routine rates have been measured for well-fed individuals ventilating in particle-free sea water, with occasional periods of probing and digging deeper into the sand (Stenton-Dozey, 1989). Active rates have been defined as corresponding to clearing and ingesting particles via the inhalent siphon, as well as digging.

For purposes of comparison, rates of oxygen consumption at the above activity levels are given for standard-sized individuals of 1 g dry tissue weight for both species (Table 4). However, it must be borne in mind that asymptotic body weights differ markedly; adult *B. digitalis* reach a maximum dry tissue weight of about 1.5 g, whereas *D. serra* may attain 5.5 g. Thus for the whelk 1 g represents a fully mature female and for the bivalve a very young, possibly just mature, male or female.

The most interesting feature emerging from Table 4 is that oxygen consumption is much higher in *Bullia* than in *Donax* for all levels of activity and regardless of whether rates are expressed for a standardized 1 g individual or as weight specific rates for asymptotic sizes. This is a reflection of the much greater mobility of the whelk. After reviewing a wide spectrum of data at different temperatures and activity levels, Bayne and Newell (1983) noted that such differences are common between predatory gastropods, where the mean value of a is 1.630, and filter-feeding molluscs, which show a mean of only 0.496.

The differences between metabolic rates at the three activity levels are, however, similar in both species. For the whelk, routine and active rates are 1.2 and 2.1 times the standard rate respectively, compared with 1.4 and 1.7 in *D. serra* (Table 4).

Brown (1981, 1982b), constructed an activity budget over a 12.4 h tidal cycle for *B. digitalis*, based on the time spent in the various activities listed in Table 4. This resulted in a mean estimate for the cost of free existence very close to the routine rate of metabolism. The budget was based on a whelk spending 88% of its time buried, a condition regarded as a standard rate in Table 4. Burial is a method of conserving energy as compared with the necessary but expensive activities associated with finding food. No such budget exists for *Donax*, but for subtidal adults at least, activity levels for the most part probably fluctuate between routine and a higher level associated with feeding and digging to maintain position in the sand.

D. *Temperature*

Acute responses of rates of oxygen consumption to short-term changes in temperature are shown graphically in Fig. 12 for three activity levels; the corresponding Q_{10} values are shown in Table 5. Both *B. digitalis* and *D. serra* (from the west coast) display a relatively flat rate-temperature curve at standard rates of metabolism (buried but ventilating). Between 10 and 20°C, temperatures which approximate ambient conditions, Q_{10} values do not exceed 1.17 for *Bullia* and 1.32 for *Donax*. Above 20°C, values are only slightly higher at 1.39 and 1.45 respectively. This relative insensitivity to temperature changes will enable both species to conserve energy as temperatures fluctuate. Such relative temperature independence during standard

A. C. BROWN, *ET AL*

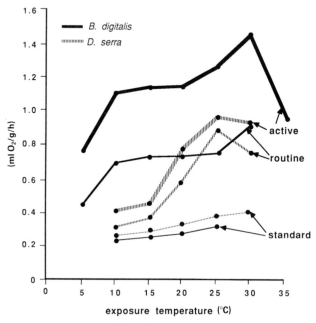

Fig. 12. Acute rate-temperature curves for *Bullia digitalis* (after Brown and da Silva, 1983) and *Donax serra* (after Stenton-Dozey, 1989), at different levels of activity.

TABLE 5. Q_{10} VALUES FOR *BULLIA DIGITALIS* AND *DONAX SERRA* HELD AT 15°C IN THE LABORATORY BEFORE BEING EXPOSED TO TEMPERATURES RANGING FROM 5 TO 30°C, FOR STANDARD, ROUTINE AND ACTIVE RATES OF METABOLISM AS DEFINED IN TABLE 4

Exposure temp. (°C)	5–10	10–15	15–20	20–25	25–30
Bullia digitalis					
Standard	—	1.17	1.16	1.39	—
Routine	2.49	1.12	1.03	1.05	1.45
Active	1.90	1.41	1.01	1.32	1.76
Donax serra					
Standard	—	1.12	1.32	1.45	1.14
Routine	—	1.28	1.61	2.30	0.73
Active	—	1.17	2.95	1.58	0.94

metabolism is found in many marine invertebrates, at least over part of the appropriate temperature range (Newell, 1979; Newell and Branch, 1981). However, that it is not a universal feature is evident from data on *Mytilus edulis* (Bayne, 1976), some species of *Patella* (Davies and Tribe, 1969) and others.

In contrast to the similarities found for standard metabolism, acute temperature responses for routine and active rates are markedly different between *B. digitalis* and *D. serra*. The routine rate of the former species is virtually independent of temperature between 10 and 25°C (Brown and da Silva, 1979), Q_{10} values varying between 1.05 and 1.12. It is positively correlated with temperature on either side of this range. The curve for active metabolism shows a similar response between 10 and 20°C but some degree of control is lost between 20 and 25°C. The Q_{10} is 1.32 compared with 1.05 for the routine rate.

Routine and active metabolic rates in *D. serra* are strongly dependent on temperature between 15 and 25°C (the Q_{10} range is 1.58 to 2.95), but less so between 10 and 15°C (Q_{10} values < 1.3). This type of dependency is the general rule among marine invertebrates (Bayne, 1976; Newell, 1979), the relative temperature independence displayed by *B. digitalis* being exceptional (Brown and da Silva, 1984).

Following longer term exposure to changed temperature conditions, *Donax* shows acclimatory adaptation (Stenton-Dozey, 1989), while *B. digitalis* does not (Brown and da Silva, 1979). For the latter species, acclimation would be little advantage since it is able to maintain a nearly constant metabolic rate over the full range of temperatures it is ever likely to encounter in the field. This is a particular advantage on the west coast, where short-term temperature fluctuations are far greater than the differences between seasonal means (see page 184). Temperature independence is not found in all species of *Bullia* and Brown and da Silva (1984) consider it a specific adaptation of *B. digitalis* to west coast conditions. Some attention has been given to the biochemical mechanisms underlying this remarkable metabolic control (da Silva, 1985; da Silva *et al.*, 1986) and it has been established that it is not due to a single feature but to the complex interactions of the properties of the enzymes concerned.

Although *Donax* shows no such conservation of energy over short-term changes in temperature for routine and active metabolic rates, acclimation after longer exposure minimizes metabolic costs, especially on the south coast, where a more gentle rise and fall in temperature is experienced. Moreover, as the animal is semi-sessile and covered by sand and water during most of the tidal cycle, it is effectively buffered from the short-term temperature fluctuations encountered on the west coast.

E. *Oxygen Availability*

Animals have been classified into two groups as far as their respiratory responses to reduced oxygen tension are concerned (Prosser, 1973), oxyconformers displaying a rate of oxygen uptake roughly proportional to ambient partial pressures, while oxyregulators maintain a level of consumption independent of declining oxygen tensions over part of the range, conforming only below a critical tension characteristic of the species. There are many examples among both gastropods and bivalves of regulators and conformers (Bayne and Newell, 1983).

The responses of the routine rates of oxygen consumpton to declining oxygen tensions (% saturation), for both *Bullia digitalis* and *Donax serra* are shown in Fig. 13. While it is clear that the bivalve is an oxyconformer, with no hint of regulatory ability (Van Wijk *et al.*, 1989), the responses of the whelk are more complex. There is actually an *increase* in the rate of oxygen uptake as ambient tensions fall from 70% to 50% saturation; this is accompanied by a marked increase in activity comparable with that involved

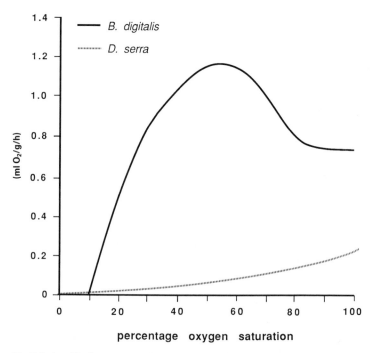

FIG. 13. Relationship between routine rate of oxygen consumption and ambient oxygen tension for *Bullia digitalis* at 10°C (after Wynberg and Brown, 1986) and for *Donax serra* at 15°C (after Van Wijk *et al.*, 1989).

in surfing (Wynberg and Brown, 1986). Since stressful conditions of all kinds stimulate responses which encourage transport by water currents (Brown, 1982c), this heightened activity may be regarded as normal behaviour. In principle, oxygen consumption is regulated down to a critical saturation of about 25%, at which level the rate of uptake is the same as under oxygen-saturated conditions. Below this tension the whelk can no longer regulate.

While it is not surprising that *Donax* is an oxyconformer in an environment where hypoxia need never be experienced, this is in marked contrast to the regulation displayed by *B. digitalis*. In addition, the bivalve incurs a considerable oxygen debt during routine metabolism at low oxygen tensions (Van Wijk *et al.*, 1989), while the whelk's rate of oxygen uptake at 100% saturation is totally unaffected by a period of hypoxia (Brown and Wynberg, 1987). The key to this difference may lie in the essentially aerobic metabolism of *Bullia* as compared with the largely anaerobic nature of metabolism in *Donax* (Trueman and Brown, 1987c).

F. *Anaerobiosis*

With few exceptions, the literature on metabolic rates in Mollusca may be criticized on the grounds that only oxygen uptake is considered as an index of those rates, the anaerobic component of metabolism being ignored. This criticism can also be levelled at the work on *Bullia* and *Donax* reviewed above, a criticism of particular weight in view of the aim of comparing the two animals. Although it seems likely that aerobiosis became important quite early in the evolution of the Mollusca, both gastropods and bivalves appear typically to have retained some anaerobic component, which can be dramatically increased under appropriate circumstances.

Bullia, indeed, appears to be somewhat exceptional in its essentially aerobic metabolism at all levels of activity. Evidence for its dependence on oxygen under normal circumstances includes the fact that all its pedal muscles are rich in mitochondria (da Silva *et al.*, 1985; da Silva and Hodgson, 1987), the finding that enough oxygen diffuses through the wall of the foot to supply the needs of the animal even when it is most active (Brown, 1984a) and the oxygen tensions measured in haemolymph from the pedal sinus, these being consistently two-thirds of ambient (Brown, 1984b, 1986b). The whelk is also unusual in that it remains expanded at all times; retraction into the shell in gastropods results in cessation of oxygen uptake and anaerobiosis is implied (Shumway, 1978). Despite its commitment to aerobic metabolism, *Bullia* is still capable of switching to anaerobiosis under conditions of extreme hypoxia in the laboratory (Brown, 1982a) and possibly also in response to changes in salinity (Brown and Meredith, 1981).

In contrast, a great deal of evidence points to considerable use of anaerobiosis in *Donax*. Trueman and Brown (1987c) recorded oxygen tensions in the pedal sinus of *D. serra* on a continuous basis and found that these did not normally exceed 50% of saturation. During burial, digging cycles commonly occurred at very low oxygen tensions, sometimes less than 10%. They also noted that the pedal musculature lacked mitochondria except at the extreme outer surface of the foot, where some oxygen presumably diffuses inwards through the pedal wall. Supporting evidence is the oxyconformity of the animal and the fact that it incurs an oxygen debt during hypoxia (Van Wijk *et al.*, 1989). Unlike *Bullia, Donax* reacts to unfavourable conditions or to danger by withdrawing totally into its shell; under these circumstances, no oxygen uptake can be measured and metabolism, though possibly greatly reduced, must be completely anaerobic.

G. *Other Factors*

While the respiratory responses of both *Donax* and *Bullia* to a wide variety of circumstances are now well documented and reviewed above, responses to changes in salinity (Brown and Meredith, 1981) and to exposure to a number of pollutants (Brown, 1982c; Brown *et al.*, 1982) have been measured only for *Bullia*. On the other hand, respiratory effects of exposure to calcium hypochlorite have been assessed only for *Donax* (Stenton-Dozey, 1989). Comparisons are therefore difficult. It does appear, however, that both animals tend to respond to adverse conditions, including pollution, by a reduction in the rate of oxygen uptake after a critical concentration has been reached. Present evidence is that this reduction is associated with a decrease in activity and a reduction in the respiratory current through the mantle cavity.

At higher concentrations of pollutants, the reactions of the two animals are quite different, resulting in markedly different respiratory responses. *Donax* tends to withdraw into its shell, as mentioned above, leading to total reliance on anaerobiosis (Stenton-Dozey, 1989), while *Bullia* does not retract but lies on the surface of the sand with its foot extended, and continues to respire aerobically (Brown, 1982c).

A few pollutants are known which, in relatively low concentration, lead to an increase rather than a decrease in oxygen consumption. Such pollutants include phenol and salts of cadmium (Golombick and Brown, 1980; Brown *et al.*, 1982). In both cases this is probably due to an uncoupling of oxidative phosphorylation.

VI. Energetics and Ecological Efficiencies

In both bivalves (Bayne and Newell, 1983; Griffiths and Griffiths, 1987) and carnivorous gastropods (Hughes, 1986), the amount of food available to an individual strongly influences each component of the energy budget, namely ingestion/consumption (C), faecal production (F), respiration (R), excretion (U) and ultimately scope for growth and reproduction (SFG), where $SFG = C - (F + R + U)$. It is thus essential, when comparing two molluscs such as *Donax* and *Bullia*, one a filter feeder, the other a carnivorous scavenger, to do so at comparable ration levels. Thus budgets for the two species are compared here at rations which maximize SFG expressed in J/day for all sizes. The inherent differences in feeding behaviour naturally necessitate different methods to estimate ingestion rates and these are addressed to make comparisons more meaningful. Data for *B. digitalis* are derived from Stenton-Dozey and Brown (1988) and those for *D. serra* from Stenton-Dozey (1989).

A. *Ingested and Absorbed Rations*

The time spent feeding, relative to the quantity of food ingested, is the most important difference to consider when comparing the ingestion rates (C) of the two species. *Bullia* is largely dependent on erratic strandings of carrion on the beach, which must be consumed quickly and in large quantities before being washed out of reach. Thus in experiments using an unlimited supply of bivalve gill tissue as food, a 1 g individual consumed 225 mg (dry weight) in 7 min, a further meal being taken only after 12 days (Stenton-Dozey and Brown, 1988). Intake was calculated over the duration of a complete feeding cycle (time feeding plus time between meals), to arrive at a daily ration of 28.28 mg dry weight of food per gram dry body weight (*a* value for ingestion in Table 6). This quantification of ingestion rate follows previous studies on the predatory gastropods *Thais lapillus* (Bayne and Scullard, 1978), *Thais carinifera* and *Natica maculosa* (Broom, 1982), in which it was recognized that the different feeding phases, namely detecting, pursuing, handling, ingesting and digesting food, constitute a feeding cycle, the duration of which equals the time denominator in quantifying ingestion. Unfortunately this approach has, as yet, not been generally adopted, so that most published data cannot be transformed into comparative values (Broom, 1982; Hughes, 1986; Stenton-Dozey and Brown, 1988).

Standard methods do exist, however, for calculating ingestion rates in suspension feeders, leading to interspecific comparable data culminating in reviews by Bayne (1976), Bayne and Newell (1983) and Griffiths and

TABLE 6. VALUES OF *a* AND *b* IN THE ALLOMETRIC EQUATION $Y = aW^b$, WHERE Y = PHYSIOLOGICAL RATE OF INGESTION (IR) AND ABSORPTION (AB) (MG DRY WT/D), PERCENTAGE BODY WT INGESTED PER DAY (IR%), RESPIRATION (R IN μG O_2/D) AND EXCRETION (U IN MG NH_4 – N/D), AND W = DRY TISSUE WT (G)

Species and diet	IR		Ab		IR %		R		U	
	a	b	a	b	a	b	a	b	a	b
Bullia digitalis (gill tissue)	28.28	0.64	24.89	0.64	2.79	−0.36	12.89	0.60	1.36	0.58
Donax serra (7.47 mg algae/l)	124.77	0.48	88.34	0.48	12.50	−0.52	10.99	0.72	0.97	0.55
Donax serra (5.46 mg detritus/l)	37.92	0.62	19.69	0.62	3.80	−0.38	7.13	0.63	0.97	0.55

Griffiths (1987). Ingestion in *D. serra* was calculated by the standard procedure of multiplying the mean change in food concentration by the relevant clearance rate (Frost, 1972). Since it is assumed that filtration is continuous and that all particles are retained with 100% efficiency, ingestion can be quantified in hourly units or as a daily ration in which periods of emergence are compensated for.

D. serra filters suspended particles from 2 μm to 20 μm in diameter from the surf water and can also retain bacteria with an 8% efficiency (Matthews *et al.*, 1989). These particles are mainly derived from offshore phytoplankton in various stages of growth or decay, as well as resuspended faeces from *D. serra* itself and from other surf organisms (Stenton-Dozey, 1989). This food source, although very variable in concentration, is constantly available to subtidal adults, whereas west coast juveniles found between LWS and MW are restricted to filtering only when covered by the tide. *D. serra* supplied with cultured algae (*Tetraselmis suecica*) at a ration level maximizing SFG (30×10^6 cells/l or 7.47 mg DW/l), ingested 124.8 mg/d/g dry tissue weight. However, when detritus in the form of stranded foam was used as food, SFG was most positive at 20×10^6 cells/l (5.46 mg DW/l), which corresponds to an ingestion rate of only 37.9 mg DW/d/g (Table 6).

These differences emphasize the necessity, for comparative purposes, of standardizing not only the methods of measuring ingestion rates but also food quality at comparable ration levels. The bivalve gill tissue used as food for *Bullia* (Stenton-Dozey and Brown, 1988) has a soft texture and represents 17.32 J/mg DW, an energy content similar to that of the cultured algae (20.7 J/mg DW) fed to *Donax*. On the other hand, 1 mg foam detritus is far more refractory than either gill tissue or the algae and is only equivalent to 9.91 J (Stenton-Dozey, 1989).

The allometric equations relating body size to ingestion rates in Table 6 indicate the similarity in rates for 1 g individuals of *Bullia* and *Donax* when the bivalve was offered detritus. On an algal diet, however, the ingestion rate was an order of magnitude higher in terms of ingested and absorbed ration (Ab in mg/d) and three times higher for weight-specific ingestion (IR% as % body wt/d) (Table 6). Consumption expressed as a percentage of 1 g dry tissue weight *Bullia* (2.79% in Table 6) falls well within the range given by Hughes (1986) for both grazing and predatory gastropods (1 to 6.2%). In comparison, *Donax* (of 1 g) held with algae ingests no less than 12.5% of its body weight/d (Table 6), which is much higher than the range of 1 to 6% recorded for other filtering bivalves under the same feeding conditions (Bayne, 1976; Winter, 1978; Navarro and Winter, 1982). However, weight-specific ingestion for *Donax* feeding on detritus does fall within the recorded range at 3.8%/d. These results could imply that in the field *Donax* has a greater potential than either its co-inhabitor, *Bullia*, or other suspension-feeding bivalves to exploit food resources of higher quality when they become available. This potential lies in an ability to accelerate clearance rates from 0.31 l/h/g, while feeding on detritus, to 0.99 l/h/g in the presence of algae, coupled with a better absorption efficiency of 71% compared to 51%. Elevated absorption efficiencies of similar percentages to those mentioned above have been observed in some mytilid species at equivalent algae/detritus ration levels (Griffiths and King, 1979; Griffiths, 1980a, b; Stuart *et al.*, 1982) but in these bivalves the increase was not associated with such a marked increase in clearance rate.

Absorption efficiencies for *D. serra* and *B. digitalis* are directly comparable, as both were calculated using the ash-ratio method (Conover, 1966). Efficiency for the whelk feeding on mussel gill (organic content 90% of the dry weight) can therefore be regarded as high, at 88%, compared to 71% for *Donax* feeding on algae (organic content 81%) and 51% when feeding on detritus (organic content 51%) at the optimal ration level. The ability of the whelk to absorb such a high proportion of food is linked to the long post-feeding phase, including digestion, which lasted up to 10 days or more in the laboratory. In the field this efficiency is clearly advantageous in the face of an erratic and unpredictable food supply and compares well with that of other carnivorous gastropods, whose measured efficiency values range from 52% to 95% (Bayne and Newell, 1983). Absorption efficiencies are in general higher in carnivorous gastropods than in suspension-feeding bivalves, since an animal food source is far less refractory and more digestible than plant material. In addition, being of non-particulate origin, it does not carry a high inorganic silt load. However, efficiency values in gastropods are more variable (Bayne and Newell, 1983), being more directly influenced by the quality rather than the quantity of food, as well as by starvation and the degree of parasitization (Calow, 1975).

The b values in Table 6 indicate the relationship between body weight (g DW of tissue) and ingestion. For *Donax* ingesting and absorbing algae, the value for b of 0.48 reveals a suppression of ingestion rates in large individuals, relative to the exponent of 0.62 on a diet of foam detritus. Other suspension-feeding bivalves have tended to show opposite trends, b values with algae as food declining from 0.66 and 0.82 to a mean of 0.45 when offered natural particulates (Bayne and Newell, 1983). However, in these instances ingestion rates (a values) were not dissimilar and hence the decline in b reflected a true suppression in large individuals. Since a values for *D. serra* held with algae are an order of magnitude higher than for detritus, it is better to view the exponent of 0.48 for an algal diet as reflecting an elevation of ingestion rates in small individuals rather than as suppression in the adults.

The weight coefficient b for *B. digitalis* (0.64 in Table 6) nearly equals that for the bivalve feeding on detritus, thus indicating parallel ingestion rates with increasing size. The only exponents in the literature that can be compared with that for *Bullia* are for predatory gastropods; these are highly variable, ranging from 0.37 to 1.01 (Bayne and Newell, 1983; Hughes, 1986). This variability is a true reflection of the diversity of time spent by predators, both intra- and inter-specifically, in different stages of feeding (searching, pursuit, capture, manipulation, ingestion); only searching and ingestion apply to scavenging in *Bullia* (Stenton-Dozey and Brown, 1988). The negative exponents for weight-specific ingestion rates in Table 6 show that food consumption in smaller whelks and bivalves is much greater in proportion to their body weight than in larger individuals. This relationship is more marked for *Donax* feeding on algae, where a 0.1 g animal can ingest 41% of its body weight per day, as compared with only 9% for the same sized mussel feeding on detritus and 6% for a whelk of the same size. This again reflects the adaptive feeding responses of *Donax serra* when faced with food potentially providing greater scope for growth and reproduction.

B. *Respiration and Excretion*

Respiration (R) in ml O_2/d and excretion (U) as mg ammonia nitrogen/d are the two measured components in the energy equation $SFG = C - (F + R + U)$, which constitute metabolic expenditure. Faecal production (F) is calculated by difference. Other unmeasured energy losses occur and, although these are frequently regarded as negligible, they are highly variable and should be borne in mind when assessing an energy budget. For instance, as mentioned in a previous section, the measure of respiration by *Bullia* and *Donax* represents only aerobic metabolism and not total metabolism. It has been shown that in some marine bivalves, anaerobiosis can account for between 5 and 7.5% of total heat loss under normoxic conditions (Bayne and Newell, 1983) and in *Mytilus edulis* this percentage may increase to 20% during starvation (Pamatmat, 1980) and 72% when emersed (Shick *et al.*, 1983). Data presented in the previous section indicate that anaerobiosis contributes much more to the total energy metabolism of *Donax* than that of *Bullia*.

Excretion (U) is conventionally measured as ammonia, this being nearly always the principal excretory product of protein catabolism in both gastropods (Hughes, 1986) and bivalves (Griffiths and Griffiths, 1987). However, in addition to ammonia, most marine molluscs also excrete amino acids, uric acid and urea, usually in small amounts but sometimes in quantities which are significant in terms of energy balance. A further parameter of U is mucous production, which may account for 9–15% of the absorbed ration or up to 40% of total production (Bayne and Newell, 1983).

Copious mucous secretion has, in fact, been witnessed during some of the activities of *Bullia* but this has not as yet been quantified. Although no measurements exist for bivalves, mucous loss appears to be less substantial, since most secretions are reingested during feeding. Thus in the final assessment it must be borne in mind that, although only ammonia excretion is considered here, other losses do occur and that losses due to mucous secretion may be quite significant in the case of the whelk.

The allometric relationships between size and respiration and excretion rates are shown in Table 6. It is apparent that for a 1 g *Donax* the large difference in ingestion rates for different diets (algal or detrital), does not apply to energy utilization. With cultured algae as food, the bivalve's respiration rate is 10.99 ml O_2 d/g while filtering at a ration level of 7.47 mg DW/l. This rate is only slightly lower, at 7.13 ml O_2 d/g, with a roughly equivalent ration of 5.46 mg detritus/l (Table 6). The difference in the weight exponents do, however, show a suppression of aerobic respiration in larger bivalves feeding on natural particulates compared to a diet of algae. It may be noted that the respiratory responses of different sized *Donax* held with a natural food source approximate those for the whelk (compare *b* values in Table 6). It must be stressed that the respiratory rates of *Bullia* presented here represent the cost of free existence and not just the rate while feeding. Nevertheless, the cost of aerobic respiration and the relationship to body size at ration levels that maximize SFG are surprisingly close for two molluscs not closely related and with such different behaviours.

Ammonia excretion rates are also very similar, both for standard sized individuals (*a* values) and with change in body size (*b* values). The rates for *Donax* show no dependence on type of diet (Table 6). This comparison can be made with confidence, as measurements for both species were made using an indophenol-blue spectrophotometric determination of ammonia (Koroleff, 1976). All the animals used were well fed; this is important as it is known that, depending on season, starvation can either accelerate ammonia excretion during protein catabolism or reduce rates when glycogen reserves are utilized (Bayne and Scullard, 1978; Stickle and Bayne, 1982). Griffiths and Griffiths (1987) have noted that measurements of ammonia excretion in bivalves are highly variable and attribute this to differences in nutritional and reproductive status of individuals. Data for gastropods are scant but exponent values appear to be far less variable, lying between 0.61 and 0.87 (Bayne and Newell, 1983), as compared with a range of 0.40 to 1.48 for bivalves (Griffiths and Griffiths, 1987).

C. Scope for Growth and Reproduction

The energy available for growth and reproduction in different sized *Bullia* and *Donax* after absorbed rations have been balanced against the cost of aerobic respiration and excretion, is shown in Fig. 14 and described in the form of allometric equations in Table 7. It is apparent that for large clams fed on algae, SFG is about four orders of magnitude higher than for animals of the same size maintained on detritus, while for small *Donax* the difference in SFG increases to 10 orders of magnitude. Algae-fed *Donax* also display about 10 times the SFG calculated for *Bullia*. These differences are consistent with differences in ingestion rates, there being no substantial differences in energy expenditure between the different diets or between the two species.

An energy budget for *Donax* fed on algae grossly overestimates SFG, even if losses accrued through mucous secretion and nitrogenous wastes other than ammonia are included in the balance. Nevertheless, as already mentioned, such a budget does provide an index of the animal's ability to maximize the intake of high energy food. The budget based on a diet of natural detritus particles is more representative of field conditions and is more appropriate for comparison with that of *Bullia*. Although SFG for the whelk is four to five times that of detritus-fed *Donax* (Table 7, Fig. 14), this may overestimate the production of *Bullia* since food in the laboratory was unlimited and of higher nutritive value than the Cnidaria which normally constitute its staple diet.

The efficiencies with which ingested (IR) and absorbed (Ab) rations are utilized for maintenance $(R + U)$ and converted into body mass (SFG = re-

TABLE 7. VALUES OF a AND b IN THE ALLOMETRIC EQUATION $Y = aW^b$, WHERE Y = DIFFERENT PHYSIOLOGICAL PROCESSES IN TERMS OF ENERGY ACQUISITION (IR AND AB IN J/D) AND ENERGY UTILIZATION (F, R, U AND SFG IN J/D) IN RELATION TO BODY WEIGHT (W IN G DRY WT) FOR *BULLIA DIGITALIS* FED BIVALVE GILL TISSUE AND FOR *DONAX SERRA* ON DIETS OF CULTURED ALGAE AND DETRITUS

Physiological process	Bullia (gills)		Donax (algae)		Donax (detritus)	
	a	b	a	b	a	b
Ingestion (IR)	486.82	0.64	2583.00	0.48	375.74	0.62
Absorption (Ab)	428.41	0.64	1829.18	0.48	191.64	0.62
Defaecation (*F*)	58.41	0.64	753.82	0.48	184.10	0.62
Respiration (*R*)	261.74	0.60	223.40	0.72	144.86	0.63
Excretion (*U*)	33.72	0.58	23.40	0.55	23.40	0.55
Growth and reproduction (SFG)	134.40	0.77	1569.30	0.45	23.26	0.64

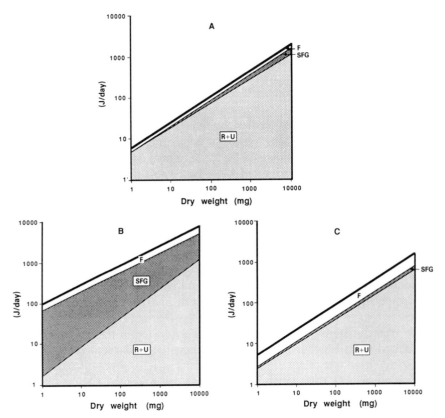

FIG. 14. Comparison of daily energy budgets for (A) *Bullia digitalis* fed on mussel gill and for *Donax serra* on diets of (B) cultured algae and (C) detritus. Energy utilization = Respiration + Excretion (R + U) and Defaecation (F). Energy available for growth and reproduction = SFG.

productive (Pr) plus somatic (Pg) growth) are shown in Fig. 15. *D. serra* utilized detritus for respiratory and excretory expenditure at efficiencies which decline with size and which are higher than on an algal diet, when efficiencies increase with size. *B. digitalis* utilizes mussel gill at efficiencies similar to those for the bivalve on a detrital diet; there is also a decline in percentages with size.

The two most commonly used indices in molluscan energetics are those of gross growth efficiency (K_1 = SFG/IR) and net growth efficiency (K_2 = SFG/Ab). Once again, for *Donax* maximum growth efficiencies are obtained on an algal rather than a detrital diet. On the former diet, K_1 values decline from 65% for a 0.1 g animal to 57% for a 5 g mussel (Fig. 15(b)). Not only are the efficiencies far less on a detrital diet but there is little difference with size, only

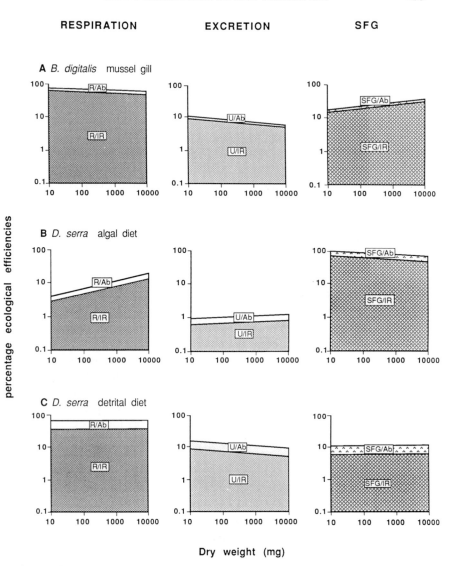

Dry weight (mg)

FIG. 15. Comparison of ecological efficiencies of respiration (R), excretion (U) and scope for growth and reproduction (SFG) as a percentage of ingestion (IR) and absorption (Ab) rates for (A) *Bullia digitalis* fed mussel gill and for *Donax serra* on diets of (B) cultured algae and (C) detritus. Note: $SFG/IR = K_1 =$ gross growth efficiency, and $SFG/Ab = K_2 =$ net growth efficiency.

a slight increase being noted (Fig. 15(c)). K_1 values for *B. digitalis*, on the other hand, increase with size from 21.4% for a 0.1 g individual to 27.3% for a whelk of 1 g (Fig. 15(a)).

Research on both bivalves (Griffiths and Griffiths, 1987) and gastropods (Hughes, 1986) has shown that gross and net growth efficiencies are usually a declining function of size, since young molluscs channel most of their energy intake into rapid somatic growth rather than into reproduction, as is the case in larger animals. Moreover, when ingestion rates increase less rapidly than respiration rates with increasing size, gross efficiencies must inevitably decline (Jørgenson, 1976). This is apparent in the data presented in Table 6: the *b* value relating size to algal ingestion rates in *Donax* (0.48) is lower than the weight exponent for respiration (0.72), so that efficiencies decrease with size. The same exponents for *Donax* feeding on detritus and *Bullia* feeding on mussel gill are very close to one another, and here growth efficiencies increase with size.

Budgets established in the laboratory need to be based on diets that resemble food resources in the field if meaningful conclusions are to be reached. Although their resources are very different for *Bullia* and *Donax* on the beach at Ou Skip, the energy each derives from its food and the way in which it is partitioned are remarkably similar. Obviously the budgets will not always compare so favourably. Periods of relative starvation will alternate with bouts of feeding, since the food supply is never consistent in either quality or quantity; these two factors have a profound effect on energy budgets. In addition, the reproductive status of the populations changes continually throughout the year, necessitating at times a greater intake of energy to maintain a positive SFG. Nevertheless, as long as energy gains outweigh losses, the gross and net growth efficiencies are likely to remain similar.

VII. Discussion

The biology of a number of invertebrate animals found on sandy beaches is well known. The majority of these inhabit shores which must be described as "sheltered" on the McLachlan scale of exposure to wave action; the sand is fine, sometimes muddy, the organic content relatively high and drainage poor, implying anoxic conditions at no great depth beneath the surface. Macrofaunal species tend to live in permanent or semi-permanent burrows. The modes of life of these species are of necessity very different from the two animals discussed in the present review. *Bullia digitalis* and *Donax serra*, although originating from very different molluscan stock, have both adapted successfully to high energy beaches, where extreme instability is the most acute problem to be overcome, permanent burrows being out of the question intertidally. Both animals are capable of powerful and rapid burrowing into the sand, a prerequisite under these unstable conditions.

They have, however, quite distinct life styles. *Bullia* is by far the more mobile, emerging from the sand to crawl and surf in search of food and responding to stress by active evasion, only rarely withdrawing into its shell. On the other hand, *D. serra*, particularly when adult, must be considered semi-sessile, maintaining a buried position in the beach for long periods of time and responding to stress by valve closure, in the manner of most but not all bivalves; as it may be noted that the free-swimming Pectinidae actively swim away from stress. It should also be stated that *D. serra*, perhaps because of its large size, is less mobile than many other species of *Donax*; it is, however, capable of emerging and migrating up and down the shore intermittently, in much the same manner as *Bullia* migrates regularly.

Bullia and *Donax* show remarkable convergence of form to the requirements of a common habitat. This is particularly well marked in respect of the foot, which functions in both as a fluid-filled cavity for the antagonism of the pedal musculature and for changing the shape of the foot to effect alternately penetration through and anchorage in the substratum. A fluid-filled cavity is a common feature of soft-bodied animals living in sand (Clark, 1964; Trueman, 1975) but is in marked contrast to the structure of the foot in rocky-shore gastropods. Indeed, perhaps the most significant morphological adaptation of *Bullia* to infaunal life is the great reduction of the pedal musculature typical of the prosobranch foot and its replacement by a fluid-filled cavity. This allows the foot to change shape readily during burial, muscle forces being transferred smoothly from one pedal region to another, and makes possible the erection of the foot as a sail during surfing. The convergence displayed by the two species in respect of burrowing even extends to the production of jets of water to facilitate movement of the shell into the sand. *Donax* follows the general pattern of burrowing bivalves in this regard, water being expelled from the mantle cavity during adduction of the valves; in *Bullia* it is brought about by compression of the mantle cavity as the shell is drawn forward by the columellar muscle.

With the exception of the Solenidae, species of *Donax* are among the most rapidly burrowing bivalves (Trueman and Brown, 1985). Similarly, comparison of the rates of burrowing in a number of gastropods, indicates that *B. digitalis* is the most powerful and rapid of the burrowing gastropods thus far investigated.

This ability on the part of both species is undoubtedly related to the nature of the habitat, in which any animal making insufficient penetration between waves is washed out of the sand and carried away. Their burrowing is thus limited to regions of the beach with saturated sand, where penetrability is reasonably high and where there is a sufficiently long interval between waves to permit adequate anchorage. Despite the marked similarities in pedal morphology and locomotory processes, several features, including the ultra-structure of the pedal musculature, are very different, underlining the fact that the similarities are due to the convergence of animals with very different ancestries.

Study of the locomotory behaviour and cost of the two species has shown the high cost of burrowing in the deeply-penetrating *D. serra* as compared with *Bullia*, which burrows only shallowly and at an oblique angle. However, as *Bullia* emerges from the sand and reburrows much more frequently than *D. serra*, the average cost of burial per tidal cycle is higher in the former, as is the cost of locomotion in general. Associated with the greater mobility of *Bullia* and its active food-searching, is a set of sensory mechanisms and responses which are much more highly developed than in *Donax*. However, neither has developed endogenous rhythms to assist the timing of tidal migrations, the animals relying entirely on physical environmental cues to time their activities.

An aspect in which the two animals have markedly different life styles concerns growth and reproduction. *D. serra* grows rapidly, attaining a shell length of some 60 mm in about five years, and could expect to live little more than a further year. *B. digitalis* reaches a shell length of 55 mm or so in about 20 years and the very oldest individuals may be nearly twice this age. Moreover, a fully grown *B. digitalis* has only about a sixth of the tissue weight of a large *D. serra*, even though the shell lengths are similar. The differences in growth rate and life span, as well as in productivity, are consistent with differences in reproductive pattern, *Donax* being a typical r-strategist, *Bullia* a k-strategist with not only a restricted breeding season and production of only one batch of offspring a year, but showing suppression of the larval stages and some maternal care of the eggs. It is interesting that both strategies appear equally successful in the extremely harsh habitat of the animals, although *Donax* generally attains a far higher biomass than *Bullia* on beaches relatively unexploited by man.

The two animals have an almost identical geographical range and favour the same beaches, these being relatively free of plant detritus, fairly gently sloping, with moderate to high (but not extremely high) exposure to wave action. In general their distribution on the beach can be related to sorting by water currents, the penetrability of the sand and, in the case of *Bullia*, to food availability. *D. serra* in addition shows partial segregation of juveniles and adults, a segregation which is not related to water currents or wave action as the zonation on the west coast is the reverse of that on the south coast. Possible determinants of this zonation are food availability, competition — including interspecific competition on the south coast, predation pressures, spat settlement patterns and temperature. In the present state of our knowledge, temperature would appear to be the most likely determinant, but this remains to be tested. Certainly *D. serra* is vulnerable to high temperatures and the adults and juveniles display differences in this regard. *B. digitalis* is less sensitive, this being consistent with the relatively long periods it spends out of the sand and its shallower burrowing. Competition, although mentioned above as a factor possibly relating to zonation, has not been discussed in this review because, despite several attempts to study it in both *Bullia* and *Donax* (Brown, unpublished), the topic remains elusive. Indeed the whole subject of competition on high energy sandy shores has been seriously neglected and is a gap in urgent need of filling.

The two animals present a marked contrast as far as respiratory physiology is concerned, central to this issue being the essentially aerobic nature of respiration in *Bullia* as opposed to considerable dependence on anaerobiosis in *Donax*, particularly during locomotion. While *Bullia* employs the surface of the foot as an important auxiliary respiratory organ, *Donax* does not; this is reflected in differences in oxygen tension in the pedal haemocoels of the animals, that in *Donax* falling to very low levels, and also in the fact that while in *Bullia* the pedal muscles are rich in mitochondria, these are absent from the foot of *Donax* except at the pedal wall. There are still further contrasts: *Bullia* is an oxyregulator, *Donax* an oxyconformer, and the latter incurs an oxygen debt, while *Bullia* does not. The rate-temperature curves are also very different, *D. serra* displaying a temperature-dependent rate of metabolism, particularly at routine and active rates, while *B. digitalis* has an almost temperature-independent rate at all levels of activity. While temperature dependence is probably unexceptional for a species of *Donax*, the temperature independence of *B. digitalis* does not extend to other species of the genus and is seen as a remarkable adaptation to the temperature fluctuations experienced on the west coast. All calculations of respiratory rates and energetic costs in *Donax* should take into account the apparently considerable contribution of anaerobiosis, so that measurements of metabolic rate conducted in a direct calorimeter constitute the most pressing need as far as future research is concerned.

Both species are opportunistic feeders, *Bullia* being exceptional in this regard, not only feeding on carrion of every description but also turning predator, growing an algal garden on its shell and making some use of dissolved organic matter. *D. serra* appears to have a greater potential than most other filter-feeders for exploiting food resources of higher quality when they become available. This potential lies in an ability to increase clearance rates substantially in the presence of high quality food. These rates are variable but overall they maximize ingestion and stimulate enhanced absorption efficiency. When the quality of suspended particles is reduced, clearance

238 A. C. BROWN, *ET AL.*

and ingestion rates decline to a low but steady level for long periods of time. This slow but continuous ingestion is accompanied by reduced absorption efficiencies. *Bullia* is adapted to consume as much as possible in one sitting, irrespective of food quality, but may increase absorption efficiency when quality is good. Adaptations of the whelk in this regard include locomotory and sensory mechanisms which enable it to find food rapidly, the use of the proboscis as a suction pump and the ability to store large amounts of food in the gut. Nevertheless, weight-specific ingestion rates reveal that small *Donax* consume far more per unit body weight than do small *Bullia*. This difference may in part account for the greater growth rates of the bivalve.

The cost of aerobic respiration in J per day and the relationship to body size at ration levels which maximize SFG are surprisingly close in the two species. On the other hand, when respiration rates are compared at equivalent activity levels (standard, routine or active), oxygen consumption in the whelk is far higher. This difference highlights the necessity of comparing energy acquisition and expenditure in balanced budgets based on food quality and quantity, rather than as isolated measurements. In conclusion, we may emphasize that the great success of both species in a common environment can be attributed not only to their individually unique morphological and physiological adaptations but also to the similar way in which they exploit very different natural resources to maintain a positive SFG.

References

Andrews, R. H. (1974). Selected aspects of upwelling research in the southern
    Benguela current. *Téthys* **6**, 327–340.
Ansell, A. D. (1981). Functional morphology and feeding of *Donax serra* Röding and
    Donax sordidus Hanley (Bivalvia: Donacidae). *Journal of Molluscan Studies* **47**,
    59–72.
Ansell, A. D. (1983). The biology of the genus *Donax. In* "Sandy Beaches as
    Ecosystems" (A. McLachlan and T. Erasmus, eds.), pp. 607–635, W. Junk, The
    Hague.
Ansell, A. D. and McLachlan, A. (1980). Upper temperature tolerances of three
    molluscs from South African sandy beaches. *Journal of Experimental Marine
    Biology and Ecology* **48**, 243–251.
Ansell, A. D. and Trevallion, A. (1969). Behavioural adaptations of intertidal
    molluscs from a tropical sandy beach. *Journal of Experimental Marine Biology and
    Ecology* **4**, 9–35.
Ansell, A. D. and Trueman, E. R. (1973). The energy cost of migration of the bivalve
    Donax on tropical sandy beaches. *Marine Behaviour and Physiology* **2**, 21–32.
Bally, R. (1981). "The ecology of three sandy beaches on the west coast of South
    Africa", PhD thesis, University of Cape Town, 404 pp.
Bally, R. (1983a). Factors affecting the distribution of organisms in the intertidal

zones of sandy beaches. *In* "Sandy Beaches as Ecosystems" (A. McLachlan and T. Erasmus, eds.), pp. 391–403, W. Junk, The Hague.

Bally, R. (1983b). Intertidal zonation on sandy beaches of the west coast of South Africa. *Cahiers de Biologie Marine* **24**, 85–103.

Bally, R. (1986). The biogeography of *Donax* (Mollusca; Bivalvia). *In* "Biology of the Genus *Donax* in Southern Africa" (T. E. Donn, ed.), pp. 7–12, University of Port Elizabeth.

Bally, R. (1987). The ecology of sandy beaches of the Benguela ecosystem. *South African Journal of Marine Science* **5**, 759–770.

Bally, R., McQuaid, C. D. and Brown, A. C. (1984). Shores of mixed sand and rock: an unexplored marine ecosystem. *South African Journal of Science* **80**, 500–503.

Bayne, B. L. (ed.) (1976). "Marine Mussels, their Ecology and Physiology", 506 pp. Cambridge University Press, Cambridge.

Bayne, B. L. and Newell, R. C. (1983). Physiological energetics of marine molluscs. *In* "The Mollusca" (A. S. M. Saleuddin and K. M. Wilbur, eds.), Vol. IV, pp. 407–523. Academic Press, New York.

Bayne, B. L. and Scullard, C. (1978). Rates of feeding by *Thais* (*Nucella*) *lapillus* (L.) *Journal of Experimental Marine Biology and Ecology* **32**, 113–129.

Birkett, D. A. (1986). Apparent fluctuations in the distribution pattern of *Donax serra* on the west coast. *In* "Biology of the genus *Donax* in Southern Africa" (T. E. Donn, ed.), pp. 30–33, University of Port Elizabeth.

Birkett, D. A. and Cook, P. A. (1987). Effect of the Benguela temperature anomaly, 1982–1983, on the breeding cycle of *Donax serra* Röding. *South African Journal of Marine Science* **5**, 191–196.

Branch, G. M. (1984). Changes in intertidal and shallow-water communities of the south and west coasts of South Africa during the 1982/1983 temperature anomaly. *South African Journal of Science* **80**, 61–65.

Branch, G. M. and Branch, M. (1981). "The Living Shores of Southern Africa." 272 pp, C. Struik, Cape Town.

Briggs, J. C. (1974). "Marine Zoogeography." 475 pp, McGraw-Hill, New York.

Broom, M. J. (1982). Size-selection, consumption rates and growth of the gastropods *Natica maculosa* (Lamarck) and *Thais carinifera* (Lamarck) preying on the bivalve *Anadara granosa* (L.). *Journal of Experimental Marine Biology and Ecology* **56**, 213–233.

Brown, A. C. (1961). Physiological–ecological studies on two sandy-beach Gastropoda from South Africa: *Bullia digitalis* Meuschen and *Bullia laevissima* (Gmelin). *Zeitschrift für Morphologie und Ökologie der Tiere* **49**, 629–657.

Brown, A. C. (1964a). Food relationships on the intertidal sandy beaches of the Cape Peninsula. *South African Journal of Science* **60**, 35–41.

Brown, A. C. (1964b). Blood volumes, blood distribution and sea-water spaces in relation to expansion and retraction of the foot in *Bullia* (Gastropoda). *Journal of Experimental Biology* **41**, 837–854.

Brown, A. C. (1971a). The ecology of the sandy beaches of the Cape Peninsula, South Africa. Part 1: Introduction. *Transactions of the Royal Society of South Africa* **39**, 247–279.

Brown, A. C. (1971b). The ecology of the sandy beaches of the Cape Peninsula, South Africa. Part 2: the mode of life of *Bullia* (Gastropoda). *Transactions of the Royal Society of South Africa* **39**, 281–333.

Brown, A. C. (1979a). The energy cost and efficiency of burrowing in the sandy-beach whelk *Bullia digitalis* (Dillwyn). *Journal of Experimental Marine Biology and Ecology* **40**, 149–154.

Brown, A. C. (1979b). The energetics of locomotion in the sandy-beach whelk *Bullia digitalis* (Dillwyn). *South African Journal of Science* **75**, 568–569.

Brown, A. C. (1979c). Oxygen consumption of the sandy-beach whelk *Bullia digitalis* Meuschen at different levels of activity. *Comparative Biochemistry and Physiology* **62A**, 673.

Brown, A. C. (1979d). Respiration and activity in *Bullia digitalis* (Dillwyn). *South African Journal of Science* **75**, 451–452.

Brown, A. C. (1981). An estimate of the cost of free existence in the sandy-beach whelk *Bullia digitalis* (Dillwyn) on the west coast of South Africa. *Journal of Experimental Marine Biology and Ecology* **49**, 51–56.

Brown, A. C. (1982a). The biology of sandy-beach whelks of the genus *Bullia* (Nassariidae). *Oceanography and Marine Biology Annual Review* **20**, 309–361.

Brown, A. C. (1982b). Towards an activity budget for the sandy-beach whelk *Bullia digitalis* (Dillwyn). *Malacologia* **22**, 681–683.

Brown, A. C. (1982c). Pollution and the sandy-beach whelk *Bullia*. *Transactions of the Royal Society of South Africa* **44**, 555–562.

Brown, A. C. (1983). The ecophysiology of sandy beach animals—a partial review. *In* "Sandy Beaches as Ecosystems" (A. McLachlan and T. Erasmus, eds.), pp. 575–605, W. Junk, The Hague.

Brown, A. C. (1984a). Oxygen diffusion through the foot of the whelk *Bullia digitalis* and its possible significance in respiration. *Journal of Experimental Marine Biology and Ecology* **79**, 1–7.

Brown, A. C. (1984b). Oxygen tensions in the pedal sinus of the whelk *Bullia digitalis* (Dillwyn). *Journal of Molluscan Studies* **50**, 122.

Brown, A. C. (1985a). The effects of crude oil pollution on marine organisms: a literature review in the South African context. *South African National Scientific Programmes Report* No. 99, 1–33.

Brown, A. C. (1985b). Egg capsules and young of *Bullia tenuis* (Nassariidae). *Journal of Molluscan Studies* **51**, 100.

Brown, A. C. (1986a). Molecular connectivity indices of organic pollutants: correlation with cessation of burrowing in the marine whelk *Bullia digitalis* (Dillwyn). *Transactions of the Royal Society of South Africa* **46**, 109–114.

Brown, A. C. (1986b). Oxygen tensions in the pedal sinus and mantle cavity of *Bullia* during hypoxia. *Journal of Molluscan Studies* **52**, 267–268.

Brown, A. C. (1987a). Marine pollution and health in South Africa. *South African Medical Journal* **71**, 244–248.

Brown, A. C. (1987b). Edible shellfish and disease in southern Africa. *Community Health in South Africa* **2** (Issue 3), 7–11.

Brown, A. C. and da Silva, F. M. (1979). The effects of temperature on oxygen consumption in *Bullia digitalis* Meuschen (Gastropoda; Prosobranchiata). *Comparative Biochemistry and Physiology* **62A**, 573–576.

Brown, A. C. and da Silva, F. M. (1983). Acute metabolic rate: temperature relationships of intact and homogenised *Bullia digitalis* (Gastropoda, Nassariidae). *Transactions of the Royal Society of South Africa* **45**, 91–96.

Brown, A. C. and da Silva, F. M. (1984). Effects of temperature on oxygen consumption in two closely-related whelks from different temperature regimes. *Journal of Experimental Marine Biology and Ecology* **84**, 145–153.

Brown, A. C. and Jarman, N. (1978). Coastal marine habitats. *In* "Biogeography and

Ecology of Southern Africa" (M. J. A. Werger, ed.), pp. 1239–1277, W. Junk, The Hague.

Brown, A. C. and Meredith, F. L. (1981). Effect of salinity changes on oxygen consumption in the sandy-beach whelk *Bullia digitalis*. *Comparative Biochemistry and Physiology* **69A**, 599–601.

Brown, A. C. and Noble, R. G. (1960). Function of the osphradium in *Bullia* (Gastropoda). *Nature* **188**, 1045.

Brown, A. C. and Trueman, E. R. (1982a). Muscles that push snails out of their shells. *Journal of Molluscan Studies* **48**, 97–98.

Brown, A. C. and Trueman, E. R. (1982b). Responses of the columellar muscle of *Bullia digitalis* to acetylcholine and to L-glutamate. *Journal of Molluscan Studies* **48**, 367–368.

Brown, A. C. and Turner, L. G. W. (1962). Expansion of the foot in *Bullia* (Gastropoda). *Nature* **195**, 98–99.

Brown, A. C. and Webb, S. C. (1985). The dark response of the whelk *Bullia digitalis* (Dillwyn). *Journal of Molluscan Studies* **51**, 351–352.

Brown, A. C. and Wynberg, R. P. (1987). Absence of an oxygen debt in the marine whelk *Bullia digitalis* (Dillwyn). *Journal of Molluscan Studies* **53**, 289–290.

Brown, A. C., Ansell, A. D. and Trevallion, A. (1978). Oxygen consumption by *Bullia* (*Dorsanum*) *melanoides* (Deshayes) and *Bullia digitalis* Meuschen (Gastropoda; Prosobranchiata)—an example of non-acclimation. *Comparative Biochemistry and Physiology* **61A**, 123–125.

Brown, A. C., Davies, K. C. and Young, D. J. (1982). Effects of cadmium and zinc on oxygen uptake in the whelk *Bullia digitalis* (Dillwyn). *Transactions of the Royal Society of South Africa* **44**, 551–554.

Brown, A. C., da Silva, F. M. and Hodgson, A. N. (1985a). Regional differentiation of the foot in a sandy-beach whelk. *Journal of Molluscan Studies* **51**, 230.

Brown, A. C., da Silva, F. M. and Orren, M. J. (1985b). Haemocyanin and protein concentrations in the blood of the sandy-beach whelk *Bullia digitalis* (Dillwyn). *Journal of Molluscan Studies* **51**, 99.

Calow, P. (1975). The feeding strategies of two freshwater gastropods, *Ancylus fluviatilis* (Müll.) and *Planorbis contortus* Linn. (Pulmonata), in terms of ingestion rates and absorption efficiencies. *Oecologia* **20**, 33–49.

Cernohorsky, W. O. (1984). Systematics of the family Nassariidae (Mollusca: Gastropoda). *Bulletin of the Auckland Institute and Museum* **14**, 1–356.

Clark, R. B. (1964). *Dynamics in Metazoan Evolution*, 313 pp., Clarendon Press, Oxford.

Colclough, J. H. and Brown, A. C. (1984). Uptake of dissolved organic matter by a marine whelk. *Transactions of the Royal Society of South Africa* **45**, 169–176.

Conover, R. J. (1966). Factors affecting the assimilation of organic matter and the question of superfluous feeding. *Limnology and Oceanography* **11**, 346–354.

Crisp, M., Davenport, J. and Shumway, S. E. (1978). Effects of feeding and of chemical stimulation on the oxygen uptake of *Nassarius reticulatus* (Gastropoda: Prosobranchia). *Journal of the Marine Biological Association of the United Kingdom* **58**, 387–399.

CSIR Pretoria (1979). The transfer of pollutants in two southern hemispheric oceanic systems. *South African National Scientific Programmes Report* No. 39, 1–185.

Cubit, J. (1969). The behaviour and physical factors causing migration and aggregation in the sand crab, *Emerita analoga* (Stimson). *Ecology* **50**, 118–123.

da Silva, F. M. (1985). "Aspects of adaptation to the environment in the whelk *Bullia.*" MSc thesis, University of Cape Town, 272 p.

da Silva, F. M. and Brown, A. C. (1984). The gardens of the sandy-beach whelk *Bullia digitalis* (Dillwyn). *Journal of Molluscan Studies* **50**, 64–65.

da Silva, F. M. and Brown, A. C. (1985). Egg capsules and veligers of the whelk *Bullia digitalis* (Gastropoda: Nassariidae). *The Veliger* **28**, 200–203.

da Silva, F. M. and Hodgson, A. N. (1987). Fine structure of the pedal muscle of the whelk *Bullia rhodostoma* Reeve: correlation with function. *Comparative Biochemistry and Physiology* **87A**, 143–149.

da Silva, F. M., Hodgson, A. N. and Brown, A. C. (1985). Vertebrate muscle characteristics in a marine invertebrate: significance for mode of life. *In* "Membranes and Muscle" (M. C. Berman *et al.*, eds.), pp. 340–341, ICSU/IRL Press, Oxford.

da Silva, F. M., Robb, F. T. and Brown, A. C. (1986). Temperature activation of foot muscle D(–)-lactate dehydrogenase in the whelk *Bullia digitalis*. *Biochimica et Biophysica Acta* **872**, 286–293.

Davies, P. S. and Tribe, N. A. (1969). Temperature dependence of metabolic rate in animals. *Nature* **224**, 723–724.

Day, J. H. (1984). "A Guide to Marine Life on South African Shores" (2nd edn.), 300 pp., A. A. Balkema, Cape Town.

Denny, M. W. (1980). Locomotion: the cost of gastropod crawling. *Science* **208**, 1288–1290.

de Villiers, G. (1975a). Reproduction of the white sand mussel *Donax serra* Röding. *Investigational Reports of the Sea Fisheries Branch of South Africa* **102**, 1–33.

de Villiers, G. (1975b). Growth, population dynamics, a mass mortality and arrangement of white sand mussels, *Donax serra* Röding, on beaches in the south-western Cape Province. *Investigational Reports of the Sea Fisheries Branch of South Africa* **109**, 1–31.

Donn, T. E. (ed.) (1986a). "Biology of the Genus *Donax* in Southern Africa." Institute for Coastal Research, University of Port Elizabeth.

Donn, T. E. (1986b). Growth, production and distributional dynamics of *Donax serra* in the Eastern Cape. *In* "Biology of the Genus *Donax* in Southern Africa" (T. E. Donn, ed.), pp. 34–41, University of Port Elizabeth.

Donn, T. E. (1987). Longshore distribution of *Donax serra* in two log-spiral bays in the eastern Cape, South Africa. *Marine Ecology Progress Series* **35**, 217–222.

Donn, T. E., Clarke, D. J., McLachlan, A. and du Toit, P. (1986). Distribution and abundance of *Donax serra* Röding (Bivalvia: Donacidae) as related to beach morphology. I. Semilunar migrations. *Journal of Experimental Marine Biology and Ecology* **102**, 121–131.

du Preez, H. H. (1984). Molluscan predation by *Ovalipes punctatus* (de Haan) (Crustacea: Brachyura: Portunidae). *Journal of Experimental Marine Biology and Ecology* **84**, 55–71.

du Preez, H. H. (1986). The white mussel, *Donax serra* Röding: food for crabs and fishes. *In* "Biology of the Genus *Donax* in Southern Africa" (T. E. Donn, ed.), pp. 41–42, University of Port Elizabeth.

Dye, A. H. and McGwynne, L. (1980). The effect of temperature and season on the respiratory rates of three psammolittoral gastropods. *Comparative Biochemistry and Physiology* **66A**, 107–111.

Ekman, S. (1953). "Zoogeography of the Sea", 417 pp., Sidgwick & Jackson, London.

Field, J. G. and Griffiths, C. L. (1989). Littoral and sublittoral ecosystems of southern

Africa. *In* "Intertidal and Littoral Ecosystems" (P. H. Niehuis and A. C. Mathieson, eds.), Elsevier, Amsterdam (in press).

Frost, B. W. (1972). Effects of size and concentration of food particles on the feeding behaviour of the marine planktonic copepod *Calanus pacificus*. *Limnology and Oceanography* **17**, 805–815.

Gilchrist, J. D. F. (1916). Observations on South African marine invertebrates. *Marine Biological Reports from South Africa* **3**, 39–47.

Golombick, T. and Brown, A. C. (1980). The effects of phenol on oxygen consumption in the whelk *Bullia digitalis* (Dillwyn). *South African Journal of Science* **76**, 375.

Griffith, E. (1833). "The Animal Kingdom, Vol. 12: Classes Mollusca and Radiata." London.

Griffiths, C. L. and Griffiths, R. J. (1987). Bivalvia. *In* "Animal Energetics" (T. J. Pandian and F. J. Vernberg, eds.), Vol. 2, pp. 1–88, Academic Press, New York.

Griffiths, C. L. and King, J. A. (1979). Some relationships between size, food availability and energy balance in the ribbed mussel *Aulacomya ater*. *Marine Biology* **51**, 141–149.

Griffiths, R. J. (1980a). Filtration, respiration and assimilation in the black mussel *Choromytilus meridionalis*. *Marine Ecology Progress Series* **3**, 63–70.

Griffiths, R. J. (1980b). Natural food availability and assimilation in the bivalve *Choromytilus meridionalis*. *Marine Ecology Progress Series* **3**, 151–156.

Harris, S. A., da Silva, F. M., Bolton, J. J. and Brown, A. C. (1986). Algal gardens and herbivory in a scavenging sandy-beach nassariid whelk. *Malacologia* **27**, 299–305.

Hart, T. J. and Currie, R. I. (1960). The Benguela Current. *Discovery Reports* **31**, 127–297.

Hazelhoff, E. F. (1938). Über die Ausnutzung des Sauerstoffs bei verschiedenen Wassertieren. *Zeitschrift für Vergleicenden Physiologie* **26**, 306–327.

Hemmingsen, A. M. (1960). Energy metabolism as related to body size and respiratory surfaces, and its evolution. *Reports of the Steno Memorial Hospital* **9**, 1–110.

Hill, A. V. (1939). The mechanical efficiency of frog muscle. *Proceedings of the Royal Society of London* **B 127**, 434–451.

Hodgson, A. N. (1982). Studies on wound healing and regeneration of the siphons of the bivalve *Donax serra* (Röding). *Transactions of the Royal Society of South Africa* **44**, 489–498.

Hodgson, A. N. (1986). Aspects of siphonal function and regeneration in *Donax*. *In* "Biology of the Genus *Donax* in Southern Africa" (T. E. Donn, ed.), pp. 7–12, University of Port Elizabeth.

Hodgson, A. N. and Brown, A. C. (1985). Contact chemoreception by the propodium of the sandy beach whelk *Bullia digitalis* (Gastropoda: Nassariidae). *Comparative Biochemistry and Physiology* **82A**, 425–427.

Hodgson, A. N. and Brown, A. C. (1987). Responses of *Bullia digitalis* (Prosobranchia, Nassariidae) to amino acids. *Journal of Molluscan Studies* **53**, 291–292.

Hodgson, A. N. and Cross, R. H. M. (1983). Studies on the structure of the foot of the genus *Bullia* (Gastropoda: Nassariidae). *Proceedings of the Electron Microscopy Society of Southern Africa* **13**, 99–100.

Hodgson, A. N. and Fielden, L. J. (1984). The structure and distribution of peripheral ciliated receptors in the bivalve mollusks *Donax serra* and *Donax sordidus*. *Journal of Molluscan Studies* **50**, 104–112.

Hughes, R. N. (1986). "A Functional Biology of Marine Gastropods", 245 pp., Croom Helm, Beckenham, Kent.

Jones, H. D. and Trueman, E. R. (1970). Locomotion of the limpet, *Patella vulgata* L. *Journal of Experimental Biology* **52**, 201–216.

Jørgensen, C. B. (1976). Growth efficiencies and factors controlling size in some mytilid bivalves, especially *Mytilus edulis* L.: a review and interpretation. *Ophelia* **15**, 175–192.

Kier, W. M. (1988). The arrangement and function of molluscan muscle. *In* "The Mollusca" (E. R. Trueman and M. R. Clarke, eds.), Vol. 11, pp. 211–252, Academic Press, New York.

Kilburn, R. and Rippey, E. (1982). "Sea Shells of Southern Africa", 249 pp, Macmillan, South Africa.

Knight, J. and Knight, R. (1986). The blood vascular system of the gills of *Pholas dactylus* L. (Mollusca, Bivalvia, Eulamellibranchia). *Philosophical Transactions of the Royal Society of London* B **313**, 509–523.

Koroleff, F. (1976). Determination of ammonia. *In* "Methods of Seawater Analysis" (K. Grasshof, ed.), pp. 126–133, Verlag Chemie, Weinhein.

Lasiak, T. A. (1983). The impact of surf-zone fish communities on infaunal assemblages associated with sandy beaches. *In* "Sandy Beaches as Ecosystems" (A. McLachlan and T. Erasmus, eds.), pp. 501–506, W. Junk, The Hague.

Lasiak, T. A. (1984). Aspects of the biology of three benthic-feeding teleosts from King's Beach, Algoa Bay. *South African Journal of Zoology* **19**, 51–56.

Matthews, S., Lucas, M. I., Stenton-Dozey, J. M. E. and Brown, A. C. (1989). Clearance and yield of bacterioplankton and particulates for two suspension-feeding, infaunal bivalves, *Donax serra* Röding and *Mactra lilacea*. *Journal of Experimental Marine Biology and Ecology* (in press).

McGwynne, L. E. (1980). "A Comparative Ecophysiological Study of Three Sandy Beach Gastropoda in the Eastern Cape." MSc thesis, University of Port Elizabeth, 143 pp.

McGwynne, L. E. (1984). Tolerances to temperature, desiccation and salinity in three sandy beach gastropods. *Comparative Biochemistry and Physiology* **79A**, 585–590.

McLachlan, A. (1977a). Studies on the psammolittoral fauna of Algoa Bay, South Africa. II. The distribution, composition and biomass of the meiofauna and macrofauna. *Zoologica Africana* **12**, 33–60.

McLachlan, A. (1977b). Composition, distribution, abundance and biomass of the macrofauna and meiofauna of four sandy beaches. *Zoologica Africana* **12**, 279–306.

McLachlan, A. (1980a). The definition of sandy beaches in relation to exposure: a simple rating system. *South African Journal of Science* **76**, 137–138.

McLachlan, A. (1980b). Intertidal zonation of the macrofauna and stratification of the meiofauna on high energy beaches in the eastern Cape, South Africa. *Transactions of the Royal Society of South Africa* **44**, 213–223.

McLachlan, A. (1981). Exposed sandy beaches as semi-closed ecosystems. *Marine Environmental Research* **4**, 59–63.

McLachlan, A. and Bate, G. (1984). Carbon budget for a high energy surf zone. *Vie et Milieu* **34**, 67–77.

McLachlan, A. and Hanekom, N. (1979). Aspects of the biology, ecology and seasonal fluctuations in biochemical composition of *Donax serra* in the East Cape. *South African Journal of Zoology* **14**, 183–193.

McLachlan, A. and Lewin, J. (1981). Observations on surf phytoplankton blooms along the coasts of South Africa. *Botanica Marina* **24**, 553–557.

McLachlan, A. and Young, N. (1982). Effects of low temperature on burrowing rates of four sandy beach molluscs. *Journal of Experimental Marine Biology and Ecology* **65**, 275–284.

McLachlan, A., Cooper, C. and van der Horst, G. (1979a). Growth and production of *Bullia rhodostoma* on an open sandy beach in Algoa Bay. *South African Journal of Zoology* **14**, 49–53.

McLachlan, A., Wooldridge, T. and van der Horst, G. (1979b). Tidal movements of the macrofauna on an exposed sandy beach in South Africa. *Journal of Zoology, London* **187**, 433–442.

Navarro, J. M. and Winter, J. E. (1982). Ingestion rate, assimilation efficiency and energy balance in *Mytilus chilensis* in relation to body size and different algal concentrations. *Marine Biology* **67**, 255–266.

Newell, P. F. (1970). Molluscs: methods for estimating production and energy flow. *In* "Methods of Study in Soil Ecology", pp. 285–291. UNESCO, Paris.

Newell, P. F. and Brown, A. C. (1977). The fine structure of the osphradium of *Bullia digitalis* Meuschen (Gastropoda; Prosobranchiata). *Malacologia* **16**, 197–205.

Newell, R. C. (1979). "Biology of Intertidal Animals" Marine Ecological Surveys, 781 pp, Faversham, Kent.

Newell, R. C. and Branch, G. M. (1981). The effects of temperature on the maintenance of metabolic energy balance in marine invertebrates. *Advances in Marine Biology* **17**, 329–396.

Pamatmat, M. M. (1980). Facultative anaerobiosis of benthos. *In* "Marine Benthic Dynamics" (E. R. Lenore and B. C. Coull, eds.), pp. 69–90, University of South Carolina Press.

Prosser, C. L. (1973). "Comparative Animal Physiology" (3rd ed.), W. B. Saunders, Philadelphia.

Runnegar, B. and Pojeta, J. Jr. (1985). Origin and diversification of the Mollusca. *In* "The Mollusca" (E. R. Trueman and M. R. Clarke, eds.), Vol. 10, pp. 1–57, Academic Press, New York.

Russell-Hunter, W. D. and Russell-Hunter, M. (1968). Pedal expansion in the naticid snails. I. Introduction and weighing experiments. *Biological Bulletin of Woods Hole* **135**, 548–562.

Shick, J. M., de Zwaan, A. and de Bont, A. M. Th. (1983). Anoxic metabolic rate in the mussel *Mytilus edulis* L. estimated by simultaneous direct calorimetry and biochemical analysis. *Physiological Zoology* **56**, 56–63.

Shillington, F. H. (1978). Surface waves near Cape Town: measurements and statistics. *Civil Engineering in South Africa* **20**, 203–206.

Shumway, S. E. (1978). The effects of fluctuating salinity on respiration in gastropod molluscs. *Comparative Biochemistry and Physiology* **63A**, 279–283.

Stanley, S. M. (1987). Adaptive morphology of the shell in bivalves and gastropods. *In* "The Mollusca" (E. R. Trueman and M. R. Clarke, eds.), Vol. 11, pp. 105–142, Academic Press, New York.

Stenton-Dozey, J. M. E. (1986). The effects of temperature and chlorination on the physiology of *Donax serra*. *In* "Biology of the Genus *Donax* in Southern Africa" (T. E. Donn, ed.), pp. 15–20. University of Port Elizabeth.

Stenton-Dozey, J. M. E. (1989). "Physiology of the Sandy Beach Bivalve *Donax serra* and Effects of Temperature and Chlorine". PhD thesis, University of Cape Town, 300 pp.

Stenton-Dozey, J. M. E. and Brown, A. C. (1988). Feeding, assimilation and scope for growth in the sandy-beach neogastropod *Bullia digitalis*. *Journal of Experimental Marine Biology and Ecology* **119**, 253–268.

Stephenson, T. A. and Stephenson, A. (1972). "Life Between Tide-marks on Rocky Shores", 425 pp, W. H. Freeman, San Francisco.

Stickle, W. B. and Bayne, B. L. (1982). Effects of temperature and salinity on oxygen consumption and nitrogen excretion in *Thais* (*Nucella*) *lapillus*. *Journal of Experimental Marine Biology and Ecology* **58**, 1–17.

Stuart, V., Field, J. G. and Newell, R. C. (1982). Evidence for absorption of kelp detritus by the ribbed mussel *Aulacomya ater* using a new ^{51}Cr-labelled microsphere technique. *Marine Ecology Progress Series* **9**, 263–271.

Trevor, J. H. (1978). The dynamics and mechanical energy expenditure of the polychaetes *Nephtys cirrosa*, *Nereis diversicolor* and *Arenicola marina* during burrowing. *Estuarine and Coastal Marine Science* **6**, 605–619.

Trueman, E. R. (1971). The control of burrowing and the migratory behaviour of *Donax denticulatus* (Bivalvia: Tellinacea). *Journal of Zoology, London* **165**, 453–467.

Trueman, E. R. (1975). "The Locomotion of Soft-bodied Animals", 200 pp, Edward Arnold, London.

Trueman, E. R. (1976). Locomotion and the origins of Mollusca. *In* "Perspectives in Experimental Biology" (P. Spencer Davies, ed.), Vol. 1, pp. 455–465, Pergamon Press, Oxford.

Trueman, E. R. (1980). Swimming by jet propulsion. *In* "Aspects of Animal Movement" (H. Y. Elder and E. R. Trueman, eds.), pp. 93–124, Cambridge University Press, Cambridge.

Trueman, E. R. (1983). Locomotion in molluscs. *In* "The Mollusca" (A. S. M. Saleuddin and K. M. Wilbur, eds.), Vol. 4, pp. 155–198, Academic Press, New York.

Trueman, E. R. and Ansell, A. D. (1969). The mechanism of burrowing into soft substrata by marine animals. *Oceanography and Marine Biology Annual Review* **7**, 315–366.

Trueman, E. R. and Brown, A. C. (1976). Locomotion, pedal retraction and extension, and the hydraulic systems of *Bullia* (Gastropoda: Nassaridae). *Journal of Zoology, London* **178**, 365–384.

Trueman, E. R. and Brown, A. C. (1985). Dynamics of burrowing and pedal extension in *Donax serra* (Mollusca: Bivalvia). *Journal of Zoology, London* **207**, 345–355.

Trueman, E. R. and Brown, A. C. (1987a). Proboscis extrusion in *Bullia* (Nassariidae): a study of fluid skeletons in Gastropoda. *Journal of Zoology, London* **211**, 505–513.

Trueman, E. R. and Brown, A. C. (1987b). Locomotory function of the pedal musculature of the nassariid whelk *Bullia*. *Journal of Molluscan Studies* **53**, 287–288.

Trueman, E. R. and Brown, A. C. (1987c). Respiration in the foot of *Donax serra*, a burrowing bivalve mollusc. *Comparative Biochemistry and Physiology* **87A**, 1059–1062.

Trueman, E. R. and Jones, H. D. (1977). Crawling and burrowing. *In* "Mechanics and Energetics of Animal Locomotion" (R. McN. Alexander and G. Goldspink, eds.), pp. 204–221, Chapman and Hall, London.

Trueman, E. R., Brand, A. R. and Davis, P. (1966). The effect of the substrate and shell shape on the burrowing of some common bivalves. *Proceedings of the Malacological Society of London* **37**, 97–109.

Trueman, E. R., Brown, A. C. and Stenton-Dozey, J. M. E. (1986). Blood flow in a burrowing bivalve at pedal extension and retraction. *Journal of Molluscan Studies* **52**, 265–266.

Tucker, V. A. (1975). The energetic cost of moving about. *American Scientist* **63**, 413–419.

Turner, H. J. and Belding, D. L. (1957). The tidal migrations of *Donax variabilis* Say. *Limnology and Oceanography* **2**, 120–124.

Underhill, L. G. and Cooper, J. (1982). "Counts of Waterbirds on the Coastline of Southern Africa", Western Cape Wader Study Group and the Percy FitzPatrick Institute of African Ornithology, Cape Town.

Van der Horst, G. (1986). *Donax serra*: reproduction as an r-strategist and potential for aquaculture. *In* "Biology of the Genus *Donax* in Southern Africa" (T. E. Donn, ed.), pp. 13–14. University of Port Elizabeth.

Van Wijk, K., Stenton-Dozey, J. M. E. and Brown, A. C. (1989). Oxyconformation in the burrowing bivalve *Donax serra* Röding. *Journal of Molluscan Studies* **55** (in press).

Voltzow, J. (1985). Morphology of the pedal circulatory system of the marine gastropod *Busycon contrarium* and its role in locomotion (Gastropoda, Buccinacea). *Zoomorphology* **105**, 395–400.

Wade, B. A. (1967). Studies on the biology of the West Indies beach clam, *Donax denticulatus* Linne. 1. Ecology. *Bulletin of Marine Science* **17**, 149–174.

Winter, J. E. (1978). A review of the knowledge of suspension-feeding in lamellibranchiate bivalves with special reference to artificial aquaculture systems. *Aquaculture* **13**, 1–33.

Wynberg, R. P. and Brown, A. C. (1986). Oxygen consumption of the whelk *Bullia digitalis* (Dillwyn) at reduced oxygen tensions. *Comparative Biochemistry and Physiology* **85A**, 45–47.

Yonge, C. M. (1947). The pallial organs in the aspidobranch Gastropoda and their evolution throughout the Mollusca. *Philosophical Transactions of the Royal Society of London* **B 230**, 79–147.

Taxonomic Index

Figures in *italic*; tables in **bold**

Subject Index

Figures in *italic*; tables in **bold**

A

Absorption efficiency, *Donax* and *Bullia*, **226**, 228, 229, 231, **231**, *233*, 238
Abundance, predation, 41–2
Acantharia
  parasites of, 125, 126, 127, 129
  symbionts, 124
Acanthocephala, parasitic, 151–2
Acoustic stimuli, larvae, predation, 16, 18–19
Actinomyxidea, parasitic, 144, 146, 158
Activity
  budgets, *Donax* and *Bullia*, 211–4, **213**, **217**, 218–9
  larvae, predation, 14, 23
Adsorption systems, parasites, 126, *128*, 131
Age, predation, 2
Aggregation, prey, 53
Algae, *Bullia*, 191, 237
Alternative prey, predation, 15, 27, 36–7, 58, 61
Ambush predation, 5, 12, 15, 23, *24*, **26**
Amine sensitivity, *Donax* and *Bullia*, 193
Amino acids, molluscs
  excretion, 229
  sensitivity, 192–3
Ammonia excretion, molluscs, 229, 230, 231
Amoebophryidae, parasitic, 126
Amphipoda
  eggs, prey, 16, 29, 43, 47
  parasites of, 135, 136, 137, 146
  parasitic, 154–6, *155*
  predators, 8, 12, 16, 36, 38
α-Amylase activity, *Donax* and *Bullia*, 192
Anaerobic metabolism, *Donax* and *Bullia*, 208, 210, 211, 213, 223–4, 229, 237

Animalia, parasitic, 146–57, *149*, *155*
Annelids, parasites of, 152
Anthomedusae, parasites of, *128*, 131
Apodinidae, parasitic, 126
Apparent Looming Threshold (ALT), predation, 16
Appendicularia, parasites of, 126, 127, 128, 129, 133, 134, 143
Aquiferous spaces, *Bullia*, 200, 205, *205*, 207
Argonautids, integument, 104
Arms, Cephalopoda, 99–100, *100*
Ascomycetes, plankton, 121
Assimilation efficiency, *Donax* and *Bullia*, 191, 192
Attachment systems, parasites, 126, *128*, 134, 135, 156
Auditory systems, larvae, predation, 16, 18–19

B

Bacteria
  diet, *Donax*, 185, 191, 227
  gut, *Bullia*, 192
  plankton, 119–20
Bacteriophages, plankton, 120
Bajkov's Equation, predation, 39, 40, 45
'Barrels', Crustacea, 155, *155*, 156, 158
Basidiomycetes, plankton, 121
Beach environment, South Africa, 183–6
Behaviourial
  factors, predation, 11
  studies, Cephalopoda, 107
Benthic invertebrates, predators, 47
Beverton-Holt Model, predation, 59
Biomass, *Donax* and *Bullia*, 180, 182, 236
Biomedical research, Cephalopoda, 107

Cumulative Index of Titles

Cumulative Index of Authors